“一带一路”建筑类大学合作与创新

Intercollegiate Cooperation and Innovation Along the Belt and Road

张爱林 主编

·北 京·

图书在版编目（CIP）数据

“一带一路”建筑类大学合作与创新/张爱林主编．—北京：国家行政管理出版社，2020. 7

ISBN 978-7-5150-2451-6

Ⅰ. ①一… Ⅱ. ①张… Ⅲ. ①“一带一路”—建筑学—高等学校—国际合作—研究 Ⅳ. ① TU ② G649. 1

中国版本图书馆 CIP 数据核字(2020)第 077027 号

书　　名 “一带一路”建筑类大学合作与创新
“YIDAIYILU” JIANZHULEI DAXUE HEZUO YU CHUANGXIN

作　　者 张爱林　主编

责任编辑 陈　科

出版发行 国家行政管理出版社
（北京市海淀区长春桥路 6 号，100089）

综 合 办 (010)68928903

发 行 部 (010)68922366　68928870

经　　销 新华书店

印　　刷 北京新视觉印刷有限公司

版　　次 2020 年 7 月北京第 1 版

印　　次 2020 年 7 月北京第 1 次印刷

开　　本 170 毫米×240 毫米　16 开

印　　张 19. 5

字　　数 299 千字

定　　价 88. 00 元

本书编委会

Editorial Board

目录

一、成立背景
Background

重要讲话 President Xi Jinping's Speech

权威发布 Official Documents

二、“一带一路”建筑类大学国际联盟概况
Overview of the Belt and Road Architectural University International Consortium

三、“一带一路”建筑类大学国际联盟大会暨校长论坛
Conferences and Presidents' Forums of the Belt and Road Architectural University International Consortium

四、校长论坛报告
Presidents' Report

一

成立背景
Background

重要讲话 President Xi Jinping′s Speech

携手推进“一带一路”建设

——在“一带一路”国际合作高峰论坛开幕式上的演讲[1]

习近平

尊敬的各位国家元首，政府首脑，
各位国际组织负责人，
女士们，先生们，朋友们：

“孟夏之日，万物并秀。”在这美好时节，来自100多个国家的各界嘉宾齐聚北京，共商“一带一路”建设合作大计，具有十分重要的意义。今天，群贤毕至，少长咸集，我期待着大家集思广益、畅所欲言，为推动“一带一路”建设献计献策，让这一世纪工程造福各国人民。

女士们、先生们、朋友们！

2000多年前，我们的先辈筚路蓝缕，穿越草原沙漠，开辟出联通亚欧非的陆上丝绸之路；我们的先辈扬帆远航，穿越惊涛骇浪，闯荡出连接东西方的海上丝绸之路。古丝绸之路打开了各国友好交往的新窗口，书写了人类发展进步的新篇章。中国陕西历史博物馆珍藏的千年“鎏金铜蚕”，在印度尼西亚发现的千年沉船“黑石号”等，见证了这段历史。

古丝绸之路绵亘万里，延续千年，积淀了以和平合作、开放包容、互学互鉴、互利共赢为核心的丝路精神。这是人类文明的宝贵遗产。

——和平合作。公元前140多年的中国汉代，一支从长安出发的和平使团，开始打通东方通往西方的道路，完成了“凿空之旅”，这就是著名的张

① 习近平：《携手推进“一带一路”建设》，新华网，2017年5月14日。

骞出使西域。中国唐宋元时期，陆上和海上丝绸之路同步发展，中国、意大利、摩洛哥的旅行家杜环、马可·波罗、伊本·白图泰都在陆上和海上丝绸之路留下了历史印记。15 世纪初的明代，中国著名航海家郑和七次远洋航海，留下千古佳话。这些开拓事业之所以名垂青史，是因为使用的不是战马和长矛，而是驼队和善意；依靠的不是坚船和利炮，而是宝船和友谊。一代又一代“丝路人”架起了东西方合作的纽带、和平的桥梁。

——开放包容。古丝绸之路跨越尼罗河流域、底格里斯河和幼发拉底河流域、印度河和恒河流域、黄河和长江流域，跨越埃及文明、巴比伦文明、印度文明、中华文明的发祥地，跨越佛教、基督教、伊斯兰教信众的汇集地，跨越不同国度和肤色人民的聚居地。不同文明、宗教、种族求同存异、开放包容，并肩书写相互尊重的壮丽诗篇，携手绘就共同发展的美好画卷。酒泉、敦煌、吐鲁番、喀什、撒马尔罕、巴格达、君士坦丁堡等古城，宁波、泉州、广州、北海、科伦坡、吉达、亚历山大等地的古港，就是记载这段历史的“活化石”。历史告诉我们：文明在开放中发展，民族在融合中共存。

——互学互鉴。古丝绸之路不仅是一条通商易货之道，更是一条知识交流之路。沿着古丝绸之路，中国将丝绸、瓷器、漆器、铁器传到西方，也为中国带来了胡椒、亚麻、香料、葡萄、石榴。沿着古丝绸之路，佛教、伊斯兰教及阿拉伯的天文、历法、医药传入中国，中国的四大发明、养蚕技术也由此传向世界。更为重要的是，商品和知识交流带来了观念创新。比如，佛教源自印度，在中国发扬光大，在东南亚得到传承。儒家文化起源中国，受到欧洲莱布尼茨、伏尔泰等思想家的推崇。这是交流的魅力、互鉴的成果。

——互利共赢。古丝绸之路见证了陆上“使者相望于道，商旅不绝于途”的盛况，也见证了海上“舶交海中，不知其数”的繁华。在这条大动脉上，资金、技术、人员等生产要素自由流动，商品、资源、成果等实现共享。阿拉木图、撒马尔罕、长安等重镇和苏尔港、广州等良港兴旺发达，罗马、安息、贵霜等古国欣欣向荣，中国汉唐迎来盛世。古丝绸之路创造了地区大发展大繁荣。

历史是最好的老师。这段历史表明，无论相隔多远，只要我们勇敢迈出第一步，坚持相向而行，就能走出一条相遇相知、共同发展之路，走向幸福

安宁和谐美好的远方。

女士们、先生们、朋友们!

从历史维度看，人类社会正处在一个大发展大变革大调整时代。世界多极化、经济全球化、社会信息化、文化多样化深入发展，和平发展的大势日益强劲，变革创新的步伐持续向前。各国之间的联系从来没有像今天这样紧密，世界人民对美好生活的向往从来没有像今天这样强烈，人类战胜困难的手段从来没有像今天这样丰富。

从现实维度看，我们正处在一个挑战频发的世界。世界经济增长需要新动力，发展需要更加普惠平衡，贫富差距鸿沟有待弥合。地区热点持续动荡，恐怖主义蔓延肆虐。和平赤字、发展赤字、治理赤字，是摆在全人类面前的严峻挑战。这是我一直在思考的问题。

2013 年秋天，我在哈萨克斯坦和印度尼西亚提出共建丝绸之路经济带和 21 世纪海上丝绸之路，即“一带一路”倡议。“桃李不言，下自成蹊。”4 年来，全球 100 多个国家和国际组织积极支持和参与“一带一路”建设，联合国大会、联合国安理会等重要决议也纳入“一带一路”建设内容。“一带一路”建设逐渐从理念转化为行动，从愿景转变为现实，建设成果丰硕。

——这是政策沟通不断深化的 4 年。我多次说过，“一带一路”建设不是另起炉灶、推倒重来，而是实现战略对接、优势互补。我们同有关国家协调政策，包括俄罗斯提出的欧亚经济联盟、东盟提出的互联互通总体规划、哈萨克斯坦提出的“光明之路”、土耳其提出的“中间走廊”、蒙古提出的“发展之路”、越南提出的“两廊一圈”、英国提出的“英格兰北方经济中心”、波兰提出的“琥珀之路”等。中国同老挝、柬埔寨、缅甸、匈牙利等国的规划对接工作也全面展开。中国同 40 多个国家和国际组织签署了合作协议，同 30 多个国家开展机制化产能合作。本次论坛期间，我们还将签署一批对接合作协议和行动计划，同 60 多个国家和国际组织共同发出推进“一带一路”贸易畅通合作倡议。各方通过政策对接，实现了“一加一大于二”的效果。

——这是设施联通不断加强的 4 年。“道路通，百业兴。”我们和相关国家一道共同加速推进雅万高铁、中老铁路、亚吉铁路、匈塞铁路等项目，建设瓜达尔港、比雷埃夫斯港等港口，规划实施一大批互联互通项目。目前，以中巴、中蒙俄、新亚欧大陆桥等经济走廊为引领，以陆海空通道和信息高速路为骨架，以铁路、港口、管网等重大工程为依托，一个复合型的基础设施网络正在形成。

——这是贸易畅通不断提升的4年。中国同“一带一路”参与国大力推动贸易和投资便利化，不断改善营商环境。我了解到，仅哈萨克斯坦等中亚国家农产品到达中国市场的通关时间就缩短了90%。2014年至2016年，中国同“一带一路”沿线国家贸易总额超过3万亿美元。中国对“一带一路”沿线国家投资累计超过500亿美元。中国企业已经在20多个国家建设56个经贸合作区，为有关国家创造近11亿美元税收和18万个就业岗位。

——这是资金融通不断扩大的4年。融资瓶颈是实现互联互通的突出挑战。中国同“一带一路”建设参与国和组织开展了多种形式的金融合作。亚洲基础设施投资银行已经为“一带一路”建设参与国的9个项目提供17亿美元贷款，“丝路基金”投资达40亿美元，中国同中东欧“16+1”金融控股公司正式成立。这些新型金融机制同世界银行等传统多边金融机构各有侧重、互为补充，形成层次清晰、初具规模的“一带一路”金融合作网络。

——这是民心相通不断促进的4年。“国之交在于民相亲，民相亲在于心相通。”“一带一路”建设参与国弘扬丝绸之路精神，开展智力丝绸之路、健康丝绸之路等建设，在科学、教育、文化、卫生、民间交往等各领域广泛开展合作，为“一带一路”建设夯实民意基础，筑牢社会根基。中国政府每年向相关国家提供1万个政府奖学金名额，地方政府也设立了丝绸之路专项奖学金，鼓励国际文教交流。各类丝绸之路文化年、旅游年、艺术节、影视桥、研讨会、智库对话等人文合作项目百花纷呈，人们往来频繁，在交流中拉近了心与心的距离。

丰硕的成果表明，“一带一路”倡议顺应时代潮流，适应发展规律，符合各国人民利益，具有广阔前景。

女士们、先生们、朋友们！

中国人说，“万事开头难”。“一带一路”建设已经迈出坚实步伐。我们要乘势而上、顺势而为，推动“一带一路”建设行稳致远，迈向更加美好的未来。这里，我谈几点意见。

第一，我们要将“一带一路”建成和平之路。古丝绸之路，和时兴，战时衰。“一带一路”建设离不开和平安宁的环境。我们要构建以合作共赢为核心的新型国际关系，打造对话不对抗、结伴不结盟的伙伴关系。各国应该尊重彼此主权、尊严、领土完整，尊重彼此发展道路和社会制度，尊重彼此核心利益和重大关切。

古丝绸之路沿线地区曾经是“流淌着牛奶与蜂蜜的地方”，如今很多地方却成了冲突动荡和危机挑战的代名词。这种状况不能再持续下去。我们要树立共同、综合、合作、可持续的安全观，营造共建共享的安全格局。要着力化解热点，坚持政治解决；要着力斡旋调解，坚持公道正义；要着力推进反恐，标本兼治，消除贫困落后和社会不公。

第二，我们要将“一带一路”建成繁荣之路。发展是解决一切问题的总钥匙。推进“一带一路”建设，要聚焦发展这个根本性问题，释放各国发展潜力，实现经济大融合、发展大联动、成果大共享。

产业是经济之本。我们要深入开展产业合作，推动各国产业发展规划相互兼容、相互促进，抓好大项目建设，加强国际产能和装备制造合作，抓住新工业革命的发展新机遇，培育新业态，保持经济增长活力。

金融是现代经济的血液。血脉通，增长才有力。我们要建立稳定、可持续、风险可控的金融保障体系，创新投资和融资模式，推广政府和社会资本合作，建设多元化融资体系和多层次资本市场，发展普惠金融，完善金融服务网络。

设施联通是合作发展的基础。我们要着力推动陆上、海上、天上、网上四位一体的联通，聚焦关键通道、关键城市、关键项目，联结陆上公路、铁路道路网络和海上港口网络。我们已经确立“一带一路”建设六大经济走廊框架，要扎扎实实向前推进。要抓住新一轮能源结构调整和能源技术变革趋势，建设全球能源互联网，实现绿色低碳发展。要完善跨区域物流网建设。我们也要促进政策、规则、标准三位一体的联通，为互联互通提供机制保障。

第三，我们要将“一带一路”建成开放之路。开放带来进步，封闭导致落后。对一个国家而言，开放如同破茧成蝶，虽会经历一时阵痛，但将换来新生。“一带一路”建设要以开放为导向，解决经济增长和平衡问题。

我们要打造开放型合作平台，维护和发展开放型世界经济，共同创造有利于开放发展的环境，推动构建公正、合理、透明的国际经贸投资规则体系，促进生产要素有序流动、资源高效配置、市场深度融合。我们欢迎各国结合自身国情，积极发展开放型经济，参与全球治理和公共产品供给，携手构建广泛的利益共同体。

贸易是经济增长的重要引擎。我们要有“向外看”的胸怀，维护多边贸易体制，推动自由贸易区建设，促进贸易和投资自由化便利化。当然，我们也要着力解决发展失衡、治理困境、数字鸿沟、分配差距等问题，建设开放、包容、普惠、平衡、共赢的经济全球化。

第四，我们要将“一带一路”建成创新之路。创新是推动发展的重要力量。“一带一路”建设本身就是一个创举，搞好“一带一路”建设也要向创新要动力。

我们要坚持创新驱动发展，加强在数字经济、人工智能、纳米技术、量子计算机等前沿领域合作，推动大数据、云计算、智慧城市建设，连接成21世纪的数字丝绸之路。我们要促进科技同产业、科技同金融深度融合，优化创新环境，集聚创新资源。我们要为互联网时代的各国青年打造创业空间、创业工场，成就未来一代的青春梦想。

我们要践行绿色发展的新理念，倡导绿色、低碳、循环、可持续的生产生活方式，加强生态环保合作，建设生态文明，共同实现2030年可持续发展目标。

第五，我们要将“一带一路”建成文明之路。“一带一路”建设要以文明交流超越文明隔阂、文明互鉴超越文明冲突、文明共存超越文明优越，推动各国相互理解、相互尊重、相互信任。

我们要建立多层次人文合作机制，搭建更多合作平台，开辟更多合作渠道。要推动教育合作，扩大互派留学生规模，提升合作办学水平。要发挥智库作用，建设好智库联盟和合作网络。在文化、体育、卫生领域，要创新合作模式，推动务实项目。要用好历史文化遗产，联合打造具有丝绸之路特色的旅游产品和遗产保护。我们要加强各国议会、政党、民间组织往来，密切妇女、青年、残疾人等群体交流，促进包容发展。我们也要加强国际反腐合作，让“一带一路”成为廉洁之路。

女士们、先生们、朋友们！

当前，中国发展正站在新的起点上。我们将深入贯彻创新、协调、绿色、开放、共享的发展理念，不断适应、把握、引领经济发展新常态，积极推进供给侧结构性改革，实现持续发展，为“一带一路”注入强大动力，为世界发展带来新的机遇。

——中国愿在和平共处五项原则基础上，发展同所有“一带一路”建设参与国的友好合作。中国愿同世界各国分享发展经验，但不会干涉他国内政，不会输出社会制度和发展模式，更不会强加于人。我们推进“一带一路”建设不会重复地缘博弈的老套路，而将开创合作共赢的新模式；不会形成破坏稳定的小集团，而将建设和谐共存的大家庭。

——中国已经同很多国家达成了“一带一路”务实合作协议，其中既包括交通运输、基础设施、能源等硬件联通项目，也包括通信、海关、检验检疫等软件联通项目，还包括经贸、产业、电子商务、海洋和绿色经济等多领域的合作规划和具体项目。中国同有关国家的铁路部门将签署深化中欧班列合作协议。我们将推动这些合作项目早日启动、早见成效。

——中国将加大对“一带一路”建设资金支持，向丝路基金新增资金1000亿元人民币，鼓励金融机构开展人民币海外基金业务，规模预计约3000亿元人民币。中国国家开发银行、进出口银行将分别提供2500亿元和1300亿元等值人民币专项贷款，用于支持“一带一路”基础设施建设、产能、金融合作。我们还将同亚洲基础设施投资银行、金砖国家新开发银行、世界银行及其他多边开发机构合作支持“一带一路”项目，同有关各方共同制定“一带一路”融资指导原则。

——中国将积极同“一带一路”建设参与国发展互利共赢的经贸伙伴关系，促进同各相关国家贸易和投资便利化，建设“一带一路”自由贸易网络，助力地区和世界经济增长。本届论坛期间，中国将同30多个国家签署经贸合作协议，同有关国家协商自由贸易协定。中国将从2018年起举办中国国际进口博览会。

——中国愿同各国加强创新合作，启动“一带一路”科技创新行动计划，开展科技人文交流、共建联合实验室、科技园区合作、技术转移4项行动。我们将在未来5年内安排2500人次青年科学家来华从事短期科研工作，培训5000人次科学技术和管理人员，投入运行50家联合实验室。我们将设立生态环保大数据服务平台，倡议建立“一带一路”绿色发展国际联盟，并为相关国家应对气候变化提供援助。

——中国将在未来3年向参与“一带一路”建设的发展中国家和国际组织提供600亿元人民币援助，建设更多民生项目。我们将向“一带一路”沿线发展中国家提供20亿元人民币紧急粮食援助，向南南合作援助基金增资10亿美元，在沿线国家实施100个“幸福家园”、100个“爱心助困”、100个“康复助医”等项目。我们将向有关国际组织提供10亿美元落实一批惠及沿线国家的合作项目。

——中国将设立“一带一路”国际合作高峰论坛后续联络机制，成立“一带一路”财经发展研究中心、“一带一路”建设促进中心，同多边开发银行共同设立多边开发融资合作中心，同国际货币基金组织合作建立能力建设中心。我们将建设丝绸之路沿线民间组织合作网络，打造新闻合作联盟、音乐教育联盟以及其他人文合作新平台。

“一带一路”建设植根于丝绸之路的历史土壤，重点面向亚欧非大陆，同时向所有朋友开放。不论来自亚洲、欧洲，还是非洲、美洲，都是“一带一路”建设国际合作的伙伴。“一带一路”建设将由大家共同商量，“一带一路”建设成果将由大家共同分享。

女士们、先生们、朋友们！

中国古语讲：“不积跬步，无以至千里。”阿拉伯谚语说，“金字塔是一块块石头垒成的”。欧洲也有句话：“伟业非一日之功”。“一带一路”建设是伟大的事业，需要伟大的实践。让我们一步一个脚印推进实施，一点一滴抓出成果，造福世界，造福人民！

祝本次高峰论坛圆满成功！

谢谢大家。

Work Together to Build the Silk Road Economic Belt and The 21st Century Maritime Silk Road[①]

Speech At the Opening Ceremony of the Belt and Road Forum for International Cooperation

Distinguished Heads of State and Government,

Heads of International Organizations,

Ladies and Gentlemen,

Dear Friends,

In this lovely season of early Summer when every living thing is full of energy, I wish to welcome all of you, distinguished guests representing over 100 countries, to attend this important forum on the Belt and Road Initiative held in Beijing. This is indeed a gathering of great minds. In the coming two days, I hope that by engaging in full exchanges of views, we will contribute to pursuing the Belt and Road Initiative, a project of the century, so that it will benefit people across the world.

Ladies and Gentlemen,

Dear Friends,

Over 2,000 years ago, our ancestors, trekking across vast steppes and deserts, opened the transcontinental passage connecting Asia, Europe and Africa,

---

习近平：*Work Together to Build the Silk Road Economic Belt and The 21st Century Maritime Silk Road*，新华网，2017 年 5 月 14 日。

known today as the Silk Road. Our ancestors, navigating rough seas, created sea routes linking the East with the West, namely, the maritime Silk Road. These ancient silk routes opened windows of friendly engagement among nations, adding a splendid chapter to the history of human progress. The thousand-year-old "gilt bronze silkworm" displayed at China's Shaanxi History Museum and the Belitung shipwreck discovered in Indonesia bear witness to this exciting period of history.

Spanning thousands of miles and years, the ancient silk routes embody the spirit of peace and cooperation, openness and inclusiveness, mutual learning and mutual benefit. The Silk Road spirit has become a great heritage of human civilization.

—Peace and cooperation. In China's Han Dynasty around 140 B. C., Zhang Qian, a royal emissary, left Chang'an, capital of the Han Dynasty. He traveled westward on a mission of peace and opened an overland route linking the East and the West, a daring undertaking which came to be known as Zhang Qian's journey to the Western regions. Centuries later, in the years of Tang, Song and Yuan Dynasties, such silk routes, both over land and at sea, boomed. Great adventurers, including Du Huan of China, Marco Polo of Italy and ibn Batutah of Morocco, left their footprints along these ancient routes. In the early 15th century, Zheng He, the famous Chinese navigator in the Ming Dynasty, made seven voyages to the Western Seas, a feat which still is remembered today. These pioneers won their place in history not as conquerors with warships, guns or swords. Rather, they are remembered as friendly emissaries leading camel caravans and sailing treasure-loaded ships. Generation after generation, the silk routes travelers have built a bridge for peace and East-West cooperation.

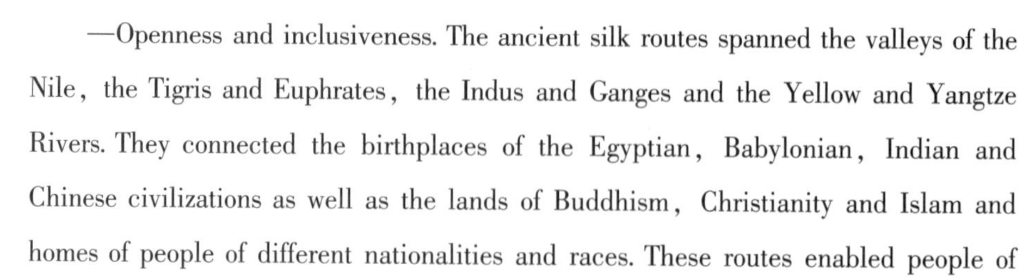

—Openness and inclusiveness. The ancient silk routes spanned the valleys of the Nile, the Tigris and Euphrates, the Indus and Ganges and the Yellow and Yangtze Rivers. They connected the birthplaces of the Egyptian, Babylonian, Indian and Chinese civilizations as well as the lands of Buddhism, Christianity and Islam and homes of people of different nationalities and races. These routes enabled people of various civilizations, religions and races to interact with and embrace each other with open mind. In the course of exchange, they fostered a spirit of mutual respect and were engaged in a common endeavor to pursue prosperity. Today, ancient cities of Jiuquan, Dunhuang, Tulufan, Kashi, Samarkand, Baghdad and Constantinople as well as ancient ports of Ningbo, Quanzhou, Guangzhou, Beihai, Colombo, Jeddah and Alexandria stand as living monuments to these past interactions. This part of history shows that civilization thrives with openness and nations prosper through exchange.

—Mutual learning. The ancient silk routes were not for trade only, they boosted flow of knowledge as well. Through these routes, Chinese silk, porcelain, lacquerwork and ironware were shipped to the West, while pepper, flax, spices, grape and pomegranate entered China. Through these routes, Buddhism, Islam and Arab astronomy, calendar and medicine found their way to China, while China's four great inventions and silkworm breeding spread to other parts of the world. More importantly, the exchange of goods and know-how spurred new ideas. For example, Buddhism originated in India, blossomed in China and was enriched in Southeast Asia. Confucianism, which was born in China, gained appreciation by European thinkers such as Leibniz and Voltaire. Herein lies the appeal of mutual learning.

—Mutual benefit. The ancient silk routes witnessed the bustling scenes of visits and trade over land and ships calling at ports. Along these major arteries of interaction, capital, technology and people flowed freely, and goods, resources and benefits were widely shared. The ancient prosperous cities of Alma-Ata, Samarkand and Chang'an and ports of Sur and Guangzhou thrived, so did the Roman Empire as well as Parthia and Kushan Kingdoms. The Han and Tang Dynasties of China entered the golden age. The ancient silk routes brought prosperity to these regions and boosted their development.

History is our best teacher. The glory of the ancient silk routes shows that geographical distance is not insurmountable. If we take the first courageous step towards each other, we can embark on a path leading to friendship, shared development, peace, harmony and a better future.

Ladies and Gentlemen,

Dear Friends,

From the historical perspective, humankind has reached an age of great progress, great transformation and profound changes. In this increasingly multi-polar, economically globalized, digitized and culturally diversified world, the trend toward peace and development becomes stronger, and reform and innovation are gaining momentum. Never have we seen such close interdependence among countries as today, such fervent desire of people for a better life, and never have we had so many means to prevail over difficulties.

In terms of reality, we find ourselves in a world fraught with challenges. Global growth requires new drivers, development needs to be more inclusive and balanced, and the gap between the rich and the poor needs to be narrowed. Hotspots in some regions are causing instability and terrorism is rampant. Deficit in peace, development and governance poses a daunting challenge to mankind. This is the issue that has always been on my mind.

In the autumn of 2013, respectively in Kazakhstan and Indonesia, I proposed the building of the Silk Road Economic Belt and the 21st Century Maritime Silk Road, which I call the Belt and Road Initiative. As a Chinese saying goes, "Peaches and plums do not speak, but they are so attractive that a path is formed below the trees." Four years on, over 100 countries and international organizations have supported and got involved in this initiative. Important resolutions passed by the UN General Assembly and Security Council contain reference to it. Thanks to our efforts, the vision of the Belt and Road Initiative is becoming a reality and bearing rich fruit.

—These four years have seen deepened policy connectivity. I have said on many occasions that the pursuit of the Belt and Road Initiative is not meant to reinvent the wheel. Rather, it aims to complement the development strategies of countries involved by leveraging their comparative strengths. We have enhanced coordination with the policy initiatives of relevant countries, such as the Eurasian Economic Union of Russia, the Master Plan on ASEAN Connectivity, the Bright Road initiative of Kazakhstan, the Middle Corridor initiative of Turkey, the Development Road initiative of Mongolia, the Two Corridors, One Economic Circle initiative of Viet Nam, the Northern Powerhouse initiative of the UK and the Amber Road initiative of Poland. We are also promoting complementarity between China's development plan and those of Laos, Cambodia, Myanmar, Hungary and other countries. China has signed cooperation agreements with over 40 countries and international organizations and carried out framework cooperation on production capacity with more than 30 countries. During the forum, a number of cooperation agreements on policy connectivity and action plans will be signed. We will also launch Belt and Road cooperation initiative on trade connectivity together with some 60 countries and international organizations. Such policy connectivity will produce a multiplying effect on cooperation among the parties involved.

—These four years have seen enhanced infrastructure connectivity. Building roads and railways creates prosperity in all sectors. We have accelerated the building of Jakarta-Bandung high-speed railway, China-Laos railway, Addis Ababa-Djibouti railway and Hungary-Serbia railway, and upgraded Gwadar and Piraeus ports in cooperation with relevant countries. A large number of connectivity projects are also in the pipeline. Today, a multi-dimensional infrastructure network is taking shape, one that is underpinned by economic corridors such as China-Pakistan Economic Corridor, China-Mongolia-Russia Economic Corridor and the New Eurasian Continental Bridge, featuring land-sea-air transportation routes and information expressway and supported by major railway, port and pipeline projects.

—These four years have seen increased trade connectivity. China has worked with other countries involved in the Belt and Road Initiative to promote trade and investment facilitation and improve business environment. I was told that for Kazakhstan and other Central Asian countries alone, customs clearance time for agricultural produce exporting to China is cut by 90%. Total trade between China and other Belt and Road countries in 2014 – 2016 has exceeded USMYM 3 trillion, and China's investment in these countries has surpassed USMYM 50 billion. Chinese companies have set up 56 economic cooperation zones in over 20 countries, generating some USMYM 1. 1 billion of tax revenue and 180,000 jobs for them.

—These four years have seen expanded financial connectivity. Financing bottleneck is a key challenge to realizing connectivity. China has engaged in multiple forms of financial cooperation with countries and organizations involved in the Belt and Road Initiative. The Asian Infrastructure Investment Bank has provided USMYM 1. 7 billion of loans for 9 projects in Belt and Road participating countries. The Silk Road Fund has made USMYM 4 billion of investment, and the "16 + 1" financial holding company between China and Central and Eastern European countries has been inaugurated. With distinctive focus, these new financial mechanisms and traditional multilateral financial institutions such as the World Bank complement each other. A multi-tiered Belt and Road financial cooperation network has taken an initial shape.

—These four years have seen strengthened people-to-people connectivity. Friendship, which derives from close contact between the people, holds the key to sound state-to-state relations. Guided by the Silk Road spirit, we the Belt and Road Initiative participating countries have pulled our efforts to build the educational Silk Road and the health Silk Road, and carried out cooperation in science, education, culture, health and people-to-people exchange. Such cooperation has helped lay a solid popular and social foundation for pursuing the Belt and Road Initiative. Every year, the Chinese government provides 10,000 government scholarships to the relevant countries. China's local governments have also set up special Silk Road scholarships to encourage international cultural and educational exchanges. Projects of people-to-people cooperation such as Silk Road culture year, tourism year, art festival, film and TV project, seminar and think tank dialogue are flourishing. These interactions have brought our people increasingly closer.

These fruitful outcomes show that the Belt and Road Initiative responds to the trend of the times, conforms to the law of development, and meets the people's interests. It surely has broad prospects.

Ladies and Gentlemen,

Dear Friends,

As we often say in China, "The beginning is the most difficult part." A solid first step has been taken in pursuing the Belt and Road Initiative. We should build on the sound momentum generated to steer the Belt and Road Initiative toward greater success. In pursuing this endeavor, we should be guided by the following principles:

First, we should build the Belt and Road into a road for peace. The ancient silk routes thrived in times of peace, but lost vigor in times of war. The pursuit of the Belt and Road Initiative requires a peaceful and stable environment. We should foster a new type of international relations featuring win-win cooperation; and we should forge partnerships of dialogue with no confrontation and of friendship rather than alliance. All countries should respect each other's sovereignty, dignity and territorial integrity, each other's development paths and social systems, and each other's core interests and major concerns.

Some regions along the ancient Silk Road used to be a land of milk and honey. Yet today, these places are often associated with conflict, turbulence, crisis and challenge. Such state of affairs should not be allowed to continue. We should foster the vision of common, comprehensive, cooperative and sustainable security, and create a security environment built and shared by all. We should work to resolve hotspot issues through political means, and promote mediation in the spirit of justice. We should intensify counter-terrorism efforts, address both its symptoms and root causes, and strive to eradicate poverty, backwardness and social injustice.

Second, we should build the Belt and Road into a road of prosperity. Development holds the master key to solving all problems. In pursuing the Belt and Road Initiative, we should focus on the fundamental issue of development, release the growth potential of various countries and achieve economic integration and interconnected development and deliver benefits to all.

Industries are the foundation of economy. We should deepen industrial cooperation so that industrial development plans of different countries will complement and reinforce each other. Focus should be put on launching major projects. We should strengthen international cooperation on production capacity and equipment manufacturing, and seize new development opportunities presented by the new industrial revolution to foster new businesses and maintain dynamic growth.

Finance is the lifeblood of modern economy. Only when the blood circulates smoothly can one grow. We should establish a stable and sustainable financial safeguard system that keeps risks under control, create new models of investment and financing, encourage greater cooperation between government and private capital and build a diversified financing system and a multi-tiered capital market. We should also develop inclusive finance and improve financial services networks.

Infrastructure connectivity is the foundation of development through cooperation. We should promote land, maritime, air and cyberspace connectivity, concentrate our efforts on key passageways, cities and projects and connect networks of highways, railways and sea ports. The goal of building six major economic corridors under the Belt and Road Initiative has been set, and we should endeavor to meet it. We need to seize opportunities presented by the new round of change in energy mix and the revolution in energy technologies to develop global energy interconnection and achieve green and low-carbon development. We should improve trans-regional logistics network and promote connectivity of policies, rules and standards so as to provide institutional safeguards for enhancing connectivity.

Third, we should build the Belt and Road into a road of opening up. Opening up brings progress while isolation results in backwardness. For a country, opening up is like the struggle of a chrysalis breaking free from its cacoon. There will be short-term pains, but such pains will create a new life. The Belt and Road Initiative should be an open one that will achieve both economic growth and balanced development.

We should build an open platform of cooperation and uphold and grow an open world economy. We should jointly create an environment that will facilitate opening up and development, establish a fair, equitable and transparent system of international trade and investment rules and boost the orderly flow of production factors, efficient resources allocation and full market integration. We welcome efforts made by other countries to grow open economies based on their national conditions, participate in global governance and provide public goods. Together, we can build a broad community of shared interests.

Trade is an important engine driving growth. We should embrace the outside world with an open mind, uphold the multilateral trading regime, advance the building of free trade areas and promote liberalization and facilitation of trade and investment. Of course, we should also focus on resolving issues such as imbalances in development, difficulties in governance, digital divide and income disparity and make economic globalization open, inclusive, balanced and beneficial to all.

Fourth, we should build the Belt and Road into a road of innovation. Innovation is an important force powering development. The Belt and Road Initiative is new by nature and we need to encourage innovation in pursuing this initiative.

We should pursue innovation-driven development and intensify cooperation in frontier areas such as digital economy, artificial intelligence, nanotechnology and quantum computing, and advance the development of big data, cloud computing and smart cities so as to turn them into a digital silk road of the 21st century. We should spur the full integration of science and technology into industries and finance, improve the environment for innovation and pool resources for innovation. We should create space and build workshops for young people of various countries to cultivate entrepreneurship in this age of the internet and help realize their dreams.

We should pursue the new vision of green development and a way of life and work that is green, low-carbon, circular and sustainable. Efforts should be made to strengthen cooperation in ecological and environmental protection and build a sound ecosystem so as to realize the goals set by the 2030 Agenda for Sustainable Development.

Fifth, we should build the Belt and Road into a road connecting different civilizations. In pursuing the Belt and Road Initiative, we should ensure that when it comes to different civilizations, exchange will replace estrangement, mutual learning will replace clashes, and coexistence will replace a sense of superiority. This will boost mutual understanding, mutual respect and mutual trust among different countries.

We should establish a multi-tiered mechanism for cultural and people-to-people exchanges, build more cooperation platforms and open more cooperation channels. Educational cooperation should be boosted, more exchange students should be encouraged and the performance of cooperatively run schools should be enhanced. Think tanks should play a better role and efforts should be made to establish think tank networks and partnerships. In the cultural, sports and health sectors, new cooperation models should be created to facilitate projects with concrete benefits. Historical and cultural heritage should be fully tapped to jointly develop tourist products and protect heritage in ways that preserve the distinctive features of the Silk Road. We should strengthen exchanges between parliaments, political parties and non-governmental organizations of different countries as well as between women, youths and people with disabilities with a view to achieving inclusive development. We should also strengthen international counter-corruption cooperation so that the Belt and Road will be a road with high ethical standards.

Ladies and Gentlemen,

Dear Friends,

China has reached a new starting point in its development endeavors. Guided by the vision of innovative, coordinated, green, open and inclusive development, we will adapt to and steer the new normal of economic development and seize opportunities it presents. We will actively promote supply-side structural reform to achieve sustainable development, inject strong impetus into the Belt and Road Initiative and create new opportunities for global development.

—China will enhance friendship and cooperation with all countries involved in the Belt and Road Initiative on the basis of the Five Principles of Peaceful Co-existence. We are ready to share practices of development with other countries, but we have no intention to interfere in other countries' internal affairs, export our own social system and model of development, or impose our own will on others. In pursuing the Belt and Road Initiative, we will not resort to outdated geopolitical maneuve-

ring. What we hope to achieve is a new model of win-win cooperation. We have no intention to form a small group detrimental to stability, what we hope to create is a big family of harmonious co-existence.

—China has reached practical cooperation agreements with many countries on pursuing the Belt and Road Initiative. These agreements cover not only projects of hardware connectivity, like transport, infrastructure and energy, but also software connectivity, involving telecommunications, customs and quarantine inspection. The agreements also include plans and projects for cooperation in economy and trade, industry, e-commerce, marine and green economy. The Chinese railway authorities will sign agreements with their counterparts of related countries to deepen cooperation on China-Europe regular railway cargo service. We will work to launch these cooperation projects at an early date and see that they deliver early benefits.

—China will scale up financing support for the Belt and Road Initiative by contributing an additional RMB 100 billion to the Silk Road Fund, and we encourage financial institutions to conduct overseas RMB fund business with an estimated amount of about RMB 300 billion. The China Development Bank and the Export-Import Bank of China will set up special lending schemes respectively worth RMB 250 billion equivalent and RMB 130 billion equivalent to support Belt and Road cooperation on infrastructure, industrial capacity and financing. We will also work with the AIIB, the BRICS New Development Bank, the World Bank and other multilateral development institutions to support Belt and Road related projects. We will work with other parties concerned to jointly formulate guidelines for financing the Belt and Road related development projects.

—China will endeavor to build a win-win business partnership with other countries participating in the Belt and Road Initiative, enhance trade and investment facilitation with them, and build a Belt and Road free trade network. These efforts are designed to promote growth both in our respective regions and globally. During this forum, China will sign business and trade cooperation agreements with over 30 countries and enter into consultation on free trade agreements with related countries. China will host the China International Import Expo starting from 2018.

—China will enhance cooperation on innovation with other countries. We will launch the Belt and Road Science, Technology and Innovation Cooperation Action Plan, which consists of the Science and Technology People-to-People Exchange Initiative, the Joint Laboratory Initiative, the Science Park Cooperation Initiative and the Technology Transfer Initiative. In the coming five years, we will offer 2,500 short-term research visits to China for young foreign scientists, train 5,000 foreign scientists, engineers and managers, and set up 50 joint laboratories. We will set up a big data service platform on ecological and environmental protection. We propose the establishment of an international coalition for green development on the Belt and Road, and we will provide support to related countries in adapting to climate change.

—In the coming three years, China will provide assistance worth RMB 60 billion to developing countries and international organizations participating in the Belt and Road Initiative to launch more projects to improve people's well-being. We will provide emergency food aid worth RMB 2 billion to developing countries along the Belt and Road and make an additional contribution of USMYM 1 billion to the Assistance Fund for South-South Cooperation. China will launch 100 "happy home" projects, 100 poverty alleviation projects and 100 health care and rehabilitation projects in countries along the Belt and Road. China will provide relevant international organizations with USMYM 1 billion to implement cooperation projects that will benefit the countries along the Belt and Road.

— China will put in place the following mechanisms to boost Belt and Road cooperation: a liaison office for the Forum's follow-up activities, the Research Center for the Belt and Road Financial and Economic Development, the Facilitating Center for Building the Belt and Road, the Multilateral Development Financial Cooperation Center in cooperation with multilateral development banks, and an IMF-China Capacity Building Center. We will also develop a network for cooperation among the NGOs in countries along the Belt and Road as well as new people-to-people exchange platforms

such as a Belt and Road news alliance and a music education alliance.

The Belt and Road Initiative is rooted in the ancient Silk Road. It focuses on the Asian, European and African continents, but is also open to all other countries. All countries, from either Asia, Europe, Africa or the Americas, can be international cooperation partners of the Belt and Road Initiative. The pursuit of this initiative is based on extensive consultation and its benefits will be shared by us all.

Ladies and Gentlemen,

Dear Friends,

An ancient Chinese saying goes, "A long journey can be covered only by taking one step at a time" . Similarly, there is an Arab proverb which says that the Pyramid was built by piling one stone on another. In Europe, there is also the saying that "Rome wasn't built in a day." The Belt and Road Initiative is a great undertaking which requires dedicated efforts. Let us pursue this initiative step by step and deliver outcome one by one. By doing so, we will bring true benefit to both the world and all our people!

In conclusion, I wish the Belt and Road Forum for International Cooperation a full success!

Thank you!

齐心开创共建“一带一路”美好未来

——在第二届“一带一路”国际合作高峰论坛开幕式上的主旨演讲①

习近平

尊敬的各位国家元首，政府首脑，

各位高级代表，

各位国际组织负责人，

女士们，先生们，朋友们：

上午好！“春秋多佳日，登高赋新诗。”在这个春意盎然的美好时节，我很高兴同各位嘉宾一道，共同出席第二届“一带一路”国际合作高峰论坛。首先，我谨代表中国政府和中国人民，并以我个人的名义，对各位来宾表示热烈的欢迎！

两年前，我们在这里举行首届高峰论坛，规划政策沟通、设施联通、贸易畅通、资金融通、民心相通的合作蓝图。今天，来自世界各地的朋友再次聚首。我期待着同大家一起，登高望远，携手前行，共同开创共建“一带一路”的美好未来。

同事们、朋友们！

共建“一带一路”倡议，目的是聚焦互联互通，深化务实合作，携手应

① 习近平：《齐心开创共建“一带一路”美好未来——在第二届“一带一路”国际合作高峰论坛开幕式上的主旨演讲》，新华网，2019年4月26日。

对人类面临的各种风险挑战，实现互利共赢、共同发展。在各方共同努力下，“六廊六路多国多港”的互联互通架构基本形成，一大批合作项目落地生根，首届高峰论坛的各项成果顺利落实，150 多个国家和国际组织同中国签署共建“一带一路”合作协议。共建“一带一路”倡议同联合国、东盟、非盟、欧盟、欧亚经济联盟等国际和地区组织的发展和合作规划对接，同各国发展战略对接。从亚欧大陆到非洲、美洲、大洋洲，共建“一带一路”为世界经济增长开辟了新空间，为国际贸易和投资搭建了新平台，为完善全球经济治理拓展了新实践，为增进各国民生福祉作出了新贡献，成为共同的机遇之路、繁荣之路。事实证明，共建“一带一路”不仅为世界各国发展提供了新机遇，也为中国开放发展开辟了新天地。

中国古人说：“万物得其本者生，百事得其道者成。”共建“一带一路”，顺应经济全球化的历史潮流，顺应全球治理体系变革的时代要求，顺应各国人民过上更好日子的强烈愿望。面向未来，我们要聚焦重点、深耕细作，共同绘制精谨细腻的“工笔画”，推动共建“一带一路”沿着高质量发展方向不断前进。

——我们要秉持共商共建共享原则，倡导多边主义，大家的事大家商量着办，推动各方各施所长、各尽所能，通过双边合作、三方合作、多边合作等各种形式，把大家的优势和潜能充分发挥出来，聚沙成塔、积水成渊。

——我们要坚持开放、绿色、廉洁理念，不搞封闭排他的小圈子，把绿色作为底色，推动绿色基础设施建设、绿色投资、绿色金融，保护好我们赖以生存的共同家园，坚持一切合作都在阳光下运作，共同以零容忍态度打击腐败。我们发起了《廉洁丝绸之路北京倡议》，愿同各方共建风清气正的丝绸之路。

——我们要努力实现高标准、惠民生、可持续目标，引入各方普遍支持的规则标准，推动企业在项目建设、运营、采购、招投标等环节按照普遍接受的国际规则标准进行，同时要尊重各国法律法规。要坚持以人民为中心的发展思想，聚焦消除贫困、增加就业、改善民生，让共建“一带一路”成果更好惠及全体人民，为当地经济社会发展作出实实在在的贡献，同时确保商业和财政上的可持续性，做到善始善终、善作善成。

同事们、朋友们！

共建“一带一路”，关键是互联互通。我们应该构建全球互联互通伙伴关系，实现共同发展繁荣。我相信，只要大家齐心协力、守望相助，即使相隔万水千山，也一定能够走出一条互利共赢的康庄大道。

基础设施是互联互通的基石，也是许多国家发展面临的瓶颈。建设高质量、可持续、抗风险、价格合理、包容可及的基础设施，有利于各国充分发挥资源禀赋，更好融入全球供应链、产业链、价值链，实现联动发展。中国将同各方继续努力，构建以新亚欧大陆桥等经济走廊为引领，以中欧班列、陆海新通道等大通道和信息高速路为骨架，以铁路、港口、管网等为依托的互联互通网络。我们将继续发挥共建“一带一路”专项贷款、丝路基金、各类专项投资基金的作用，发展丝路主题债券，支持多边开发融资合作中心有效运作。我们欢迎多边和各国金融机构参与共建“一带一路”投融资，鼓励开展第三方市场合作，通过多方参与实现共同受益的目标。

商品、资金、技术、人员流通，可以为经济增长提供强劲动力和广阔空间。“河海不择细流，故能就其深。”如果人为阻断江河的流入，再大的海，迟早都有干涸的一天。我们要促进贸易和投资自由化便利化，旗帜鲜明反对保护主义，推动经济全球化朝着更加开放、包容、普惠、平衡、共赢的方向发展。我们将同更多国家商签高标准自由贸易协定，加强海关、税收、审计监管等领域合作，建立共建“一带一路”税收征管合作机制，加快推广“经认证的经营者”国际互认合作。我们还制定了《“一带一路”融资指导原则》，发布了《“一带一路”债务可持续性分析框架》，为共建“一带一路”融资合作提供指南。中方今年将举办第二届中国国际进口博览会，为各方进入中国市场搭建更广阔平台。

创新就是生产力，企业赖之以强，国家赖之以盛。我们要顺应第四次工业革命发展趋势，共同把握数字化、网络化、智能化发展机遇，共同探索新技术、新业态、新模式，探寻新的增长动能和发展路径，建设数字丝绸之路、创新丝绸之路。中国将继续实施共建“一带一路”科技创新行动计划，同各方一道推进科技人文交流、共建联合实验室、科技园区合作、技术转移四大举措。我们将积极实施创新人才交流项目，未来5年支持5000人次中外方创

新人才开展交流、培训、合作研究。我们还将支持各国企业合作推进信息通信基础设施建设，提升网络互联互通水平。

发展不平衡是当今世界最大的不平衡。在共建“一带一路”过程中，要始终从发展的视角看问题，将可持续发展理念融入项目选择、实施、管理的方方面面。我们要致力于加强国际发展合作，为发展中国家营造更多发展机遇和空间，帮助他们摆脱贫困，实现可持续发展。为此，我们同各方共建“一带一路”可持续城市联盟、绿色发展国际联盟，制定《“一带一路”绿色投资原则》，发起“关爱儿童、共享发展，促进可持续发展目标实现”合作倡议。我们启动共建“一带一路”生态环保大数据服务平台，将继续实施绿色丝路使者计划，并同有关国家一道，实施“一带一路”应对气候变化南南合作计划。我们还将深化农业、卫生、减灾、水资源等领域合作，同联合国在发展领域加强合作，努力缩小发展差距。

我们要积极架设不同文明互学互鉴的桥梁，深入开展教育、科学、文化、体育、旅游、卫生、考古等各领域人文合作，加强议会、政党、民间组织往来，密切妇女、青年、残疾人等群体交流，形成多元互动的人文交流格局。未来5年，中国将邀请共建“一带一路”国家的政党、智库、民间组织等1万名代表来华交流。我们将鼓励和支持沿线国家社会组织广泛开展民生合作，联合开展一系列环保、反腐败等领域培训项目，深化各领域人力资源开发合作。我们将持续实施“丝绸之路”中国政府奖学金项目，举办“一带一路”青年创意与遗产论坛、青年学生“汉语桥”夏令营等活动。我们还将设立共建“一带一路”国际智库合作委员会、新闻合作联盟等机制，汇聚各方智慧和力量。

同事们、朋友们！

今年是中华人民共和国成立70周年。70年前，中国人民历经几代人上下求索，终于在中国共产党领导下建立了新中国，中国人民从此站了起来，中国人民的命运从此掌握在了自己手中。

历经70年艰苦奋斗，中国人民立足本国国情，在实践中不断探索前进方向，开辟了中国特色社会主义道路。今天的中国，已经站在新的历史起点上。我们深知，尽管成就辉煌，但前方还有一座座山峰需要翻越，还有一个个险滩等待跋涉。我们将继续沿着中国特色社会主义道路大步向前，坚持全面深化改革，坚持高质量发展，坚持扩大对外开放，坚持走和平发展道路，推动构建人类命运共同体。

下一步，中国将采取一系列重大改革开放举措，加强制度性、结构性安排，促进更高水平对外开放。

第一，更广领域扩大外资市场准入。公平竞争能够提高效率、带来繁荣。中国已实施准入前国民待遇加负面清单管理模式，未来将继续大幅缩减负面清单，推动现代服务业、制造业、农业全方位对外开放，并在更多领域允许外资控股或独资经营。我们将新布局一批自由贸易试验区，加快探索建设自由贸易港。我们将加快制定配套法规，确保严格实施《外商投资法》。我们将以公平竞争、开放合作推动国内供给侧结构性改革，有效淘汰落后和过剩产能，提高供给体系质量和效率。

第二，更大力度加强知识产权保护国际合作。没有创新就没有进步。加强知识产权保护，不仅是维护内外资企业合法权益的需要，更是推进创新型国家建设、推动高质量发展的内在要求。中国将着力营造尊重知识价值的营商环境，全面完善知识产权保护法律体系，大力强化执法，加强对外国知识产权人合法权益的保护，杜绝强制技术转让，完善商业秘密保护，依法严厉打击知识产权侵权行为。中国愿同世界各国加强知识产权保护合作，创造良好创新生态环境，推动同各国在市场化法治化原则基础上开展技术交流合作。

第三，更大规模增加商品和服务进口。中国既是“世界工厂”，也是“世界市场”。中国有世界上规模最大、成长最快的中等收入群体，消费增长潜力巨大。为满足人民日益增长的物质文化生活需要，增加消费者选择和福利，我们将进一步降低关税水平，消除各种非关税壁垒，不断开大中国市场大门，欢迎来自世界各国的高质量产品。我们不刻意追求贸易顺差，愿意进口更多国外有竞争力的优质农产品、制成品和服务，促进贸易平衡发展。

第四，更加有效实施国际宏观经济政策协调。全球化的经济需要全球化的治理。中国将加强同世界各主要经济体的宏观政策协调，努力创造正面外溢效应，共同促进世界经济强劲、可持续、平衡、包容增长。中国不搞以邻为壑的汇率贬值，将不断完善人民币汇率形成机制，使市场在资源配置中起决定性作用，保持人民币汇率在合理均衡水平上的基本稳定，促进世界经济稳定。规则和信用是国际治理体系有效运转的基石，也是国际经贸关系发展的前提。中国积极支持和参与世贸组织改革，共同构建更高水平的国际经贸规则。

第五，更加重视对外开放政策贯彻落实。中国人历来讲求“一诺千金”。我们高度重视履行同各国达成的多边和双边经贸协议，加强法治政府、诚信政府建设，建立有约束的国际协议履约执行机制，按照扩大开放的需要修改完善法律法规，在行政许可、市场监管等方面规范各级政府行为，清理废除妨碍公平竞争、扭曲市场的不合理规定、补贴和做法，公平对待所有企业和经营者，完善市场化、法治化、便利化的营商环境。

中国扩大开放的举措，是根据中国改革发展客观需要作出的自主选择，这有利于推动经济高质量发展，有利于满足人民对美好生活的向往，有利于世界和平、稳定、发展。我们也希望世界各国创造良好投资环境，平等对待中国企业、留学生和学者，为他们正常开展国际交流合作活动提供公平友善的环境。我们坚信，一个更加开放的中国，将同世界形成更加良性的互动，带来更加进步和繁荣的中国和世界。

同事们、朋友们！

让我们携起手来，一起播撒合作的种子，共同收获发展的果实，让各国人民更加幸福，让世界更加美好！

祝本次高峰论坛圆满成功！

谢谢大家。

Working Together to Deliver a Brighter Future for Belt and Road Cooperation[①]

Speech At the Opening Ceremony of the Second Belt and Road Forum for International Cooperation

Your Excellencies Heads of State and Government,

Your Excellencies High-level Representatives,

Your Excellencies Heads of International Organizations,

Ladies and Gentlemen,

Friends,

Good morning! As a line of a classical Chinese poem goes, "Spring and autumn are lovely seasons in which friends get together to climb up mountains and write poems." On this beautiful spring day, it gives me great pleasure to have you with us here at the Second Belt and Road Forum for International Cooperation (BRF). On behalf of the Chinese government and people and in my own name, I extend a very warm welcome to you all!

Two years ago, it was here that we met for the First Belt and Road Forum for International Cooperation, where we drew a blueprint of cooperation to enhance policy, infrastructure, trade, financial and people-to-people connectivity. Today, we

---

① 习近平：*Working Together to Deliver a Brighter Future for Belt and Road Cooperation*，新华网，2019年4月27日。

are once again meeting here with you, friends from across the world. I look forward to scaling new heights with you and enhancing our partnership. Together, we will create an even brighter future for Belt and Road cooperation.

Dear Colleagues and Friends,

The joint pursuit of the Belt and Road Initiative (BRI) aims to enhance connectivity and practical cooperation. It is about jointly meeting various challenges and risks confronting mankind and delivering win-win outcomes and common development. Thanks to the joint efforts of all of us involved in this initiative, a general connectivity framework consisting of six corridors, six connectivity routes and multiple countries and ports has been put in place. A large number of cooperation projects have been launched, and the decisions of the first BRF have been smoothly implemented. More than 150 countries and international organizations have signed agreements on Belt and Road cooperation with China. The complementarity between the BRI and the development plans or cooperation initiatives of international and regional organizations such as the United Nations, the Association of Southeast Asian Nations, the African Union, the European Union, the Eurasian Economic Union and between the BRI and the development strategies of the participating countries has been enhanced. From the Eurasian continent to Africa, the Americas and Oceania, Belt and Road cooperation has opened up new space for global economic growth, produced new platforms for international trade and investment and offered new ways for improving global economic governance. Indeed, this initiative has helped improve people's lives in countries involved and created more opportunities for common prosperity. What we have achieved amply demonstrates that Belt and Road cooperation has both generated new opportunities for the development of all participating countries and opened up new horizon for China's development and opening-up.

An ancient Chinese philosopher observed that "plants with strong roots grow well, and efforts with the right focus will ensure success." The Belt and Road cooperation embraces the historical trend of economic globalization, responds to the call for improving the global governance system and meets people's longing for a better life. Going ahead, we should focus on priorities and project execution, move forward with results-oriented implementation, just like an architect refining the blueprint, and jointly promote high-quality Belt and Road cooperation.

—We need to be guided by the principle of extensive consultation, joint contribution and shared benefits. We need to act in the spirit of multilateralism, pursue cooperation through consultation and keep all participants motivated. We may, by engaging in bilateral, trilateral and multilateral cooperation, fully tap into the strengths of all participants. Just as a Chinese proverb says, "A tower is built when soil on earth accumulates, and a river is formed when streams come together."

—We need to pursue open, green and clean cooperation. The Belt and Road is not an exclusive club; it aims to promote green development. We may launch green infrastructure projects, make green investment and provide green financing to protect the Earth which we all call home. In pursuing Belt and Road cooperation, everything should be done in a transparent way, and we should have zero tolerance for corruption. The Beijing Initiative for Clean Silk Road has been launched, which represents our strong commitment to transparency and clean governance in pursuing Belt and Road cooperation.

—We need to pursue high standard cooperation to improve people's lives and promote sustainable development. We will adopt widely accepted rules and standards and encourage participating companies to follow general international rules and standards in project development, operation, procurement and tendering and bidding. The laws and regulations of participating countries should also be respected. We need to take a people-centered approach, give priority to poverty alleviation and job creation to see that the joint pursuit of Belt and Road cooperation will deliver true benefits to the people of participating countries and contribute to their social and economic development. We also need to ensure the commercial and fiscal sustainability of all projects so that they will achieve the intended goals as planned.

Dear Colleagues and Friends,

Connectivity is vital to advancing Belt and Road cooperation. We need to pro-

mote a global partnership of connectivity to achieve common development and prosperity. I am confident that as we work closely together, we will transcend geographical distance and embark on a path of win-win cooperation.

Infrastructure is the bedrock of connectivity, while the lack of infrastructure has held up the development of many countries. High-quality, sustainable, resilient, affordable, inclusive and accessible infrastructure projects can help countries fully leverage their resource endowment, better integrate into the global supply, industrial and value chains, and realize inter-connected development. To this end, China will continue to work with other parties to build a connectivity network centering on economic corridors such as the New Eurasian Land Bridge, supplemented by major transportation routes like the China-Europe Railway Express and the New International Land-Sea Trade Corridor and information expressway, and reinforced by major railway, port and pipeline projects. We will continue to make good use of the Belt and Road Special Lending Scheme, the Silk Road Fund, and various special investment funds, develop Silk Road theme bonds, and support the Multilateral Cooperation Center for Development Finance in its operation. We welcome the participation of multilateral and national financial institutions in BRI investment and financing and encourage third-market cooperation. With the involvement of multiple stakeholders, we can surely deliver benefits to all.

The flow of goods, capital, technology and people will power economic growth and create broad space for it. As a Chinese saying goes, "The ceaseless inflow of rivers makes the ocean deep." However, were such inflow to be cut, the ocean, however big, would eventually dry up. We need to promote trade and investment liberalization and facilitation, say no to protectionism, and make economic globalization more open, inclusive, balanced and beneficial to all. To this end, we will enter into negotiation with more countries to conclude high-standard free trade agreements, and strengthen cooperation in customs, taxation and audit oversight by setting up the Belt and Road Initiative Tax Administration Cooperation Mechanism and accelerating international collaboration on the mutual recognition of Authorized Economic Operators. We have also formulated the Guiding Principles on Financing the Development of the Belt and Road and published the Debt Sustainability Framework for Participating Countries of the Belt and Road Initiative to provide guidance for BRI financing cooperation. In addition, the Second China International Import Expo will be held this year to build an even bigger platform for other parties to access the Chinese market.

Innovation boosts productivity; it makes companies competitive and countries strong. We need to keep up with the trend of the Fourth Industrial Revolution, jointly seize opportunities created by digital, networked and smart development, explore new technologies and new forms and models of business, foster new growth drivers and explore new development pathways, and build the digital Silk Road and the Silk Road of innovation. China will continue to carry out the Belt and Road Science, Technology and Innovation Cooperation Action Plan, and will work with our partners to pursue four major initiatives, namely the Science and Technology People-to-People Exchange Initiative, the Joint Laboratory Initiative, the Science Park Cooperation Initiative, and the Technology Transfer Initiative. We will actively implement the Belt and Road Initiative Talents Exchange Program, and will, in the coming five years, offer 5,000 opportunities for exchange, training and cooperative research for talents from China and other BRI participating countries. We will also support companies of various countries in jointly advancing ICT infrastructure building to upgrade cyber connectivity.

Imbalance in development is the greatest imbalance confronting today's world. In the joint pursuit of the BRI, we must always take a development-oriented approach and see that the vision of sustainable development underpins project selection, implementation and management. We need to strengthen international development cooperation so as to create more opportunities for developing countries, help them eradicate poverty and achieve sustainable development. In this connection, China and its partners have set up the Belt and Road Sustainable Cities Alliance and the BRI International Green Development Coalition, formulated the Green Investment Principles for the Belt and Road Development, and launched the Declaration on Accelerating the Sustainable Development Goals for Children through Shared Development. We have set up the BRI Environmental Big Data Platform. We will continue to implement the Green Silk Road Envoys Program and work with relevant countries to jointly implement the Belt and Road South-South Cooperation Initiative on Climate Change. We will also deepen cooperation in agriculture, health, disaster mitigation and water resources; and we will enhance development cooperation with the United Nations to narrow the gap in development.

We need to build bridges for exchanges and mutual learning among different cultures, deepen cooperation in education, science, culture, sports, tourism, health and archaeology, strengthen exchanges between parliaments, political parties and non-governmental organizations and exchanges between women, young people and people with disabilities in order to facilitate multi-faceted people-to-people exchanges. To this end, we will, in the coming five years, invite 10,000 representatives of political parties, think tanks and non-governmental organizations from Belt and Road participating countries to visit China. We will encourage and support extensive cooperation on livelihood projects among social organizations of participating countries, conduct a number of environmental protection and anti-corruption training courses and deepen human resources development cooperation in various areas. We will continue to run the Chinese government scholarship Silk Road Program, host the International Youth Forum on Creativity and Heritage along the Silk Roads and the "Chinese Bridge" summer camps. We will also put in place new mechanisms such as the Belt and Road Studies Network and the Belt and Road News Alliance to draw inspiration and pool our strength for greater synergy.

Dear Colleagues and Friends,

This year marks the 70th anniversary of the founding of the People's Republic of China. Seven decades ago, through the arduous struggle carried out by several generations of Chinese people and under the leadership of the Communist Party of China, New China was founded. We Chinese have since stood up and held our future in our own hands.

Over the past seven decades, we in China have, based on its realities, constantly explored the way forward through practices, and have succeeded in following the path of socialism with Chinese characteristics. Today, China has reached a new historical starting point. However, we are keenly aware that with all we have achieved, there are still many mountains to scale and many shoals to navigate. We will continue to advance along the path of socialism with Chinese characteristics, deepen sweeping reforms, pursue quality development, and expand opening-up. We remain committed to peaceful development and will endeavor to build a community with a shared future for mankind.

Going forward, China will take a series of major reform and opening-up measures and make stronger institutional and structural moves to boost higher quality opening-up.

First, we will expand market access for foreign investment in more areas. Fair competition boosts business performance and creates prosperity. China has already adopted a management model based on pre-establishment national treatment and negative list, and will continue to significantly shorten the negative list. We will work for the all-round opening-up of modern services, manufacturing and agriculture, and will allow the operation of foreign-controlled or wholly foreign-owned businesses in more sectors. We will plan new pilot free trade zones and explore at a faster pace the opening of a free trade port. We will accelerate the adoption of supporting regulations to ensure full implementation of the Foreign Investment Law. We will promote supply-side structural reform through fair competition and open cooperation, and will phase out backward and excessive production capacity in an effective way to improve the quality and efficiency of supply.

Second, we will intensify efforts to enhance international cooperation in intellectual property protection. Without innovation, there will be no progress. Full intellectual property protection will not only ensure the lawful rights and interests of Chinese and foreign companies; it is also crucial to promoting China's innovation-driven

and quality development. China will spare no effort to foster a business environment that respects the value of knowledge. We will fully improve the legal framework for protecting intellectual property, step up law enforcement, enhance protection of the lawful rights and interests of foreign intellectual property owners, stop forced technology transfer, improve protection of trade secrets, and crack down hard on violations of intellectual property in accordance with law. China will strengthen cooperation with other countries in intellectual property protection, create an enabling environment for innovation and promote technological exchanges and cooperation with other countries on the basis of market principles and the rule of law.

Third, we will increase the import of goods and services on an even larger scale. China is both a global factory and a global market. With the world's largest and fastest growing middle-income population, China has a vast potential for increasing consumption. To meet our people's ever-growing material and cultural needs and give our consumers more choices and benefits, we will further lower tariffs and remove various non-tariff barriers. We will steadily open China's market wider to quality products from across the world. China does not seek trade surplus; we want to import more competitive quality agricultural products, manufactured goods and services to promote balanced trade.

Fourth, we will more effectively engage in international macro-economic policy coordination. A globalized economy calls for global governance. China will strengthen macro policy coordination with other major economies to generate a positive spillover and jointly contribute to robust, sustainable, balanced and inclusive global growth. China will not resort to the beggar-thy-neighbor practice of RMB devaluation. On the contrary, we will continue to improve the exchange rate regime, see that the market plays a decisive role in resource allocation and keep the RMB exchange rate basically stable at an adaptive and equilibrium level. These steps will help ensure the steady growth of the global economy. Rules and credibility underpin the effective functioning of the international governance system; they are the prerequisite for growing international economic and trade relations. China is an active supporter and participant of WTO reform and will work with others to develop international economic and trade rules of higher standards.

Fifth, we will work harder to ensure the implementation of opening-up related policies. We Chinese have a saying that honoring a promise carries the weight of gold. We are committed to implementing multilateral and bilateral economic and trade agreements reached with other countries. We will strengthen the building of a government based on the rule of law and good faith. A binding mechanism for honoring international agreements will be put in place. Laws and regulations will be revised and improved in keeping with the need to expand opening-up. We will see that governments at all levels operate in a well-regulated way when it comes to issuing administrative licenses and conducting market oversight. We will overhaul and abolish unjustified regulations, subsidies and practices that impede fair competition and distort the market. We will treat all enterprises and business entities equally, and foster an enabling business environment based on market operation and governed by law.

These measures to expand opening-up are a choice China has made by itself to advance its reform and development. It will promote high-quality economic development, meet the people's desire for a better life, and contribute to world peace, stability and development. We hope that other countries will also create an enabling environment of investment, treat Chinese enterprises, students and scholars as equals, and provide a fair and friendly environment for them to engage in international exchanges and cooperation. We are convinced that a more open China will further integrate itself into the world and deliver greater progress and prosperity for both China and the world at large.

Dear Colleagues and Friends,

Let us join hands to sow the seeds of cooperation, harvest the fruits of development, bring greater happiness to our people and make our world a better place for all!

In conclusion, I wish the Second Belt and Road Forum for International Cooperation a full success!

Thank you!

权威发布 Official Documents

推动共建丝绸之路经济带和21世纪海上丝绸之路的愿景与行动

国家发展改革委　外交部　商务部

2015年3月

前　言

2000多年前，亚欧大陆上勤劳勇敢的人民，探索出多条连接亚欧非几大文明的贸易和人文交流通路，后人将其统称为“丝绸之路”。千百年来，“和平合作、开放包容、互学互鉴、互利共赢”的丝绸之路精神薪火相传，推进了人类文明进步，是促进沿线各国繁荣发展的重要纽带，是东西方交流合作的象征，是世界各国共有的历史文化遗产。

进入21世纪，在以和平、发展、合作、共赢为主题的新时代，面对复苏乏力的全球经济形势，纷繁复杂的国际和地区局面，传承和弘扬丝绸之路精神更显重要和珍贵。

2013年9月和10月，中国国家主席习近平在出访中亚和东南亚国家期间，先后提出共建“丝绸之路经济带”和“21世纪海上丝绸之路”（以下简称“一带一路”）的重大倡议，得到国际社会高度关注。中国国务院总理李

资料来源：《推动共建丝绸之路经济带和21世纪海上丝绸之路的愿景与行动》，人民网，2017年4月。

克强参加2013年中国—东盟博览会时强调，铺就面向东盟的海上丝绸之路，打造带动腹地发展的战略支点。加快“一带一路”建设，有利于促进沿线各国经济繁荣与区域经济合作，加强不同文明交流互鉴，促进世界和平发展，是一项造福世界各国人民的伟大事业。

“一带一路”建设是一项系统工程，要坚持共商、共建、共享原则，积极推进沿线国家发展战略的相互对接。为推进实施“一带一路”重大倡议，让古丝绸之路焕发新的生机活力，以新的形式使亚欧非各国联系更加紧密，互利合作迈向新的历史高度，中国政府特制定并发布《推动共建丝绸之路经济带和21世纪海上丝绸之路的愿景与行动》。

一、时代背景

当今世界正发生复杂深刻的变化，国际金融危机深层次影响继续显现，世界经济缓慢复苏、发展分化，国际投资贸易格局和多边投资贸易规则酝酿深刻调整，各国面临的发展问题依然严峻。共建“一带一路”顺应世界多极化、经济全球化、文化多样化、社会信息化的潮流，秉持开放的区域合作精神，致力于维护全球自由贸易体系和开放型世界经济。共建“一带一路”旨在促进经济要素有序自由流动、资源高效配置和市场深度融合，推动沿线各国实现经济政策协调，开展更大范围、更高水平、更深层次的区域合作，共同打造开放、包容、均衡、普惠的区域经济合作架构。共建“一带一路”符合国际社会的根本利益，彰显人类社会共同理想和美好追求，是国际合作以及全球治理新模式的积极探索，将为世界和平发展增添新的正能量。

共建“一带一路”致力于亚欧非大陆及附近海洋的互联互通，建立和加强沿线各国互联互通伙伴关系，构建全方位、多层次、复合型的互联互通网络，实现沿线各国多元、自主、平衡、可持续的发展。“一带一路”的互联互通项目将推动沿线各国发展战略的对接与耦合，发掘区域内市场的潜力，促进投资和消费，创造需求和就业，增进沿线各国人民的人文交流与文明互鉴，让各国人民相逢相知、互信互敬，共享和谐、安宁、富裕的生活。

当前，中国经济和世界经济高度关联。中国将一以贯之地坚持对外开放的基本国策，构建全方位开放新格局，深度融入世界经济体系。推进“一带一路”建设既是中国扩大和深化对外开放的需要，也是加强和亚欧非及世界各国互利合作的需要，中国愿意在力所能及的范围内承担更多责任义务，为人类和平发展作出更大的贡献。

二、共建原则

恪守联合国宪章的宗旨和原则。遵守和平共处五项原则，即尊重各国主权和领土完整、互不侵犯、互不干涉内政、和平共处、平等互利。

坚持开放合作。“一带一路”相关的国家基于但不限于古代丝绸之路的范围，各国和国际、地区组织均可参与，让共建成果惠及更广泛的区域。

坚持和谐包容。倡导文明宽容，尊重各国发展道路和模式的选择，加强不同文明之间的对话，求同存异、兼容并蓄、和平共处、共生共荣。

坚持市场运作。遵循市场规律和国际通行规则，充分发挥市场在资源配置中的决定性作用和各类企业的主体作用，同时发挥好政府的作用。

坚持互利共赢。兼顾各方利益和关切，寻求利益契合点和合作最大公约数，体现各方智慧和创意，各施所长，各尽所能，把各方优势和潜力充分发挥出来。

三、框架思路

“一带一路”是促进共同发展、实现共同繁荣的合作共赢之路，是增进理解信任、加强全方位交流的和平友谊之路。中国政府倡议，秉持和平合作、开放包容、互学互鉴、互利共赢的理念，全方位推进务实合作，打造政治互信、经济融合、文化包容的利益共同体、命运共同体和责任共同体。

“一带一路”贯穿亚欧非大陆，一头是活跃的东亚经济圈，一头是发达的欧洲经济圈，中间广大腹地国家经济发展潜力巨大。丝绸之路经济带重点畅通中国经中亚、俄罗斯至欧洲（波罗的海）；中国经中亚、西亚至波斯湾、地中海；中国至东南亚、南亚、印度洋。21 世纪海上丝绸之路重点方向是从中国沿海港口过南海到印度洋，延伸至欧洲；从中国沿海港口过南海到南太平洋。

根据“一带一路”走向，陆上依托国际大通道，以沿线中心城市为支撑，以重点经贸产业园区为合作平台，共同打造新亚欧大陆桥、中蒙俄、中国—中亚—西亚、中国—中南半岛等国际经济合作走廊；海上以重点港口为节点，共同建设通畅安全高效的运输大通道。中巴、孟中印缅两个经济走廊与推进“一带一路”建设关联紧密，要进一步推动合作，取得更大进展。

“一带一路”建设是沿线各国开放合作的宏大经济愿景，需各国携手努力，朝着互利互惠、共同安全的目标相向而行。努力实现区域基础设施更加完善，安全高效的陆海空通道网络基本形成，互联互通达到新水平；投资贸易便利化水平进一步提升，高标准自由贸易区网络基本形成，经济联系更加紧密，政治互信更加深入；人文交流更加广泛深入，不同文明互鉴共荣，各国人民相知相交、和平友好。

四、合作重点

沿线各国资源禀赋各异，经济互补性较强，彼此合作潜力和空间很大。以政策沟通、设施联通、贸易畅通、资金融通、民心相通为主要内容，重点在以下方面加强合作。

政策沟通。加强政策沟通是“一带一路”建设的重要保障。加强政府间合作，积极构建多层次政府间宏观政策沟通交流机制，深化利益融合，促进政治互信，达成合作新共识。沿线各国可以就经济发展战略和对策进行充分交流对接，共同制定推进区域合作的规划和措施，协商解决合作中的问题，共同为务实合作及大型项目实施提供政策支持。

设施联通。基础设施互联互通是“一带一路”建设的优先领域。在尊重相关国家主权和安全关切的基础上，沿线国家宜加强基础设施建设规划、技术标准体系的对接，共同推进国际骨干通道建设，逐步形成连接亚洲各次区域以及亚欧非之间的基础设施网络。强化基础设施绿色低碳化建设和运营管理，在建设中充分考虑气候变化影响。

抓住交通基础设施的关键通道、关键节点和重点工程，优先打通缺失路

段，畅通瓶颈路段，配套完善道路安全防护设施和交通管理设施设备，提升道路通达水平。推进建立统一的全程运输协调机制，促进国际通关、换装、多式联运有机衔接，逐步形成兼容规范的运输规则，实现国际运输便利化。推动口岸基础设施建设，畅通陆水联运通道，推进港口合作建设，增加海上航线和班次，加强海上物流信息化合作。拓展建立民航全面合作的平台和机制，加快提升航空基础设施水平。

加强能源基础设施互联互通合作，共同维护输油、输气管道等运输通道安全，推进跨境电力与输电通道建设，积极开展区域电网升级改造合作。

共同推进跨境光缆等通信干线网络建设，提高国际通信互联互通水平，畅通信息丝绸之路。加快推进双边跨境光缆等建设，规划建设洲际海底光缆项目，完善空中（卫星）信息通道，扩大信息交流与合作。

贸易畅通。投资贸易合作是“一带一路”建设的重点内容。宜着力研究解决投资贸易便利化问题，消除投资和贸易壁垒，构建区域内和各国良好的营商环境，积极同沿线国家和地区共同商建自由贸易区，激发释放合作潜力，做大做好合作“蛋糕”。

沿线国家宜加强信息互换、监管互认、执法互助的海关合作，以及检验检疫、认证认可、标准计量、统计信息等方面的双多边合作，推动世界贸易组织《贸易便利化协定》生效和实施。改善边境口岸通关设施条件，加快边境口岸“单一窗口”建设，降低通关成本，提升通关能力。加强供应链安全与便利化合作，推进跨境监管程序协调，推动检验检疫证书国际互联网核查，开展“经认证的经营者”（AEO）互认。降低非关税壁垒，共同提高技术性贸易措施透明度，提高贸易自由化便利化水平。

拓宽贸易领域，优化贸易结构，挖掘贸易新增长点，促进贸易平衡。创新贸易方式，发展跨境电子商务等新的商业业态。建立健全服务贸易促进体系，巩固和扩大传统贸易，大力发展现代服务贸易。把投资和贸易有机结合起来，以投资带动贸易发展。

加快投资便利化进程，消除投资壁垒。加强双边投资保护协定、避免双重征税协定磋商，保护投资者的合法权益。

拓展相互投资领域，开展农林牧渔业、农机及农产品生产加工等领域深度合作，积极推进海水养殖、远洋渔业、水产品加工、海水淡化、海洋生物制药、海洋工程技术、环保产业和海上旅游等领域合作。加大煤炭、油气、金属矿产等传统能源资源勘探开发合作，积极推动水电、核电、风电、太阳能等清洁、可再生能源合作，推进能源资源就地就近加工转化合作，形成能源资源合作上下游一体化产业链。加强能源资源深加工技术、装备与工程服务合作。

推动新兴产业合作，按照优势互补、互利共赢的原则，促进沿线国家加强在新一代信息技术、生物、新能源、新材料等新兴产业领域的深入合作，推动建立创业投资合作机制。

优化产业链分工布局，推动上下游产业链和关联产业协同发展，鼓励建立研发、生产和营销体系，提升区域产业配套能力和综合竞争力。扩大服务业相互开放，推动区域服务业加快发展。探索投资合作新模式，鼓励合作建设境外经贸合作区、跨境经济合作区等各类产业园区，促进产业集群发展。在投资贸易中突出生态文明理念，加强生态环境、生物多样性和应对气候变化合作，共建绿色丝绸之路。

中国欢迎各国企业来华投资。鼓励本国企业参与沿线国家基础设施建设和产业投资。促进企业按属地化原则经营管理，积极帮助当地发展经济、增加就业、改善民生，主动承担社会责任，严格保护生物多样性和生态环境。

资金融通。资金融通是“一带一路”建设的重要支撑。深化金融合作，推进亚洲货币稳定体系、投融资体系和信用体系建设。扩大沿线国家双边本币互换、结算的范围和规模。推动亚洲债券市场的开放和发展。共同推进亚洲基础设施投资银行、金砖国家开发银行筹建，有关各方就建立上海合作组织融资机构开展磋商。加快丝路基金组建运营。深化中国—东盟银行联合体、上合组织银行联合体务实合作，以银团贷款、银行授信等方式开展多边金融合作。支持沿线国家政府和信用等级较高的企业以及金融机构在中国境内发行人民币债券。符合条件的中国境内金融机构和企业可以在境外发行人民币债券和外币债券，鼓励在沿线国家使用所筹资金。

加强金融监管合作，推动签署双边监管合作谅解备忘录，逐步在区域内建立高效监管协调机制。完善风险应对和危机处置制度安排，构建区域性金融风险预警系统，形成应对跨境风险和危机处置的交流合作机制。加强征信

管理部门、征信机构和评级机构之间的跨境交流与合作。充分发挥丝路基金以及各国主权基金作用，引导商业性股权投资基金和社会资金共同参与“一带一路”重点项目建设。

民心相通。民心相通是“一带一路”建设的社会根基。传承和弘扬丝绸之路友好合作精神，广泛开展文化交流、学术往来、人才交流合作、媒体合作、青年和妇女交往、志愿者服务等，为深化双多边合作奠定坚实的民意基础。

扩大相互间留学生规模，开展合作办学，中国每年向沿线国家提供1万个政府奖学金名额。沿线国家间互办文化年、艺术节、电影节、电视周和图书展等活动，合作开展广播影视剧精品创作及翻译，联合申请世界文化遗产，共同开展世界遗产的联合保护工作。深化沿线国家间人才交流合作。

加强旅游合作，扩大旅游规模，互办旅游推广周、宣传月等活动，联合打造具有丝绸之路特色的国际精品旅游线路和旅游产品，提高沿线各国游客签证便利化水平。推动21世纪海上丝绸之路邮轮旅游合作。积极开展体育交流活动，支持沿线国家申办重大国际体育赛事。

强化与周边国家在传染病疫情信息沟通、防治技术交流、专业人才培养等方面的合作，提高合作处理突发公共卫生事件的能力。为有关国家提供医疗援助和应急医疗救助，在妇幼健康、残疾人康复以及艾滋病、结核、疟疾等主要传染病领域开展务实合作，扩大在传统医药领域的合作。

加强科技合作，共建联合实验室（研究中心）、国际技术转移中心、海上合作中心，促进科技人员交流，合作开展重大科技攻关，共同提升科技创新能力。

整合现有资源，积极开拓和推进与沿线国家在青年就业、创业培训、职业技能开发、社会保障管理服务、公共行政管理等共同关心领域的务实合作。

充分发挥政党、议会交往的桥梁作用，加强沿线国家之间立法机构、主要党派和政治组织的友好往来。开展城市交流合作，欢迎沿线国家重要城市之间互结友好城市，以人文交流为重点，突出务实合作，形成更多鲜活的合作范例。欢迎沿线国家智库之间开展联合研究、合作举办论坛等。

加强沿线国家民间组织的交流合作，重点面向基层民众，广泛开展教育医疗、减贫开发、生物多样性和生态环保等各类公益慈善活动，促进沿线贫困地区生产生活条件改善。加强文化传媒的国际交流合作，积极利用网络平台，运用新媒体工具，塑造和谐友好的文化生态和舆论环境。

五、合作机制

当前，世界经济融合加速发展，区域合作方兴未艾。积极利用现有双多边合作机制，推动“一带一路”建设，促进区域合作蓬勃发展。

加强双边合作，开展多层次、多渠道沟通磋商，推动双边关系全面发展。推动签署合作备忘录或合作规划，建设一批双边合作示范。建立完善双边联合工作机制，研究推进“一带一路”建设的实施方案、行动路线图。充分发挥现有联委会、混委会、协委会、指导委员会、管理委员会等双边机制作用，协调推动合作项目实施。

强化多边合作机制作用，发挥上海合作组织（SCO）、中国—东盟“10+1”、亚太经合组织（APEC）、亚欧会议（ASEM）、亚洲合作对话（ACD）、亚信会议（CICA）、中阿合作论坛、中国—海合会战略对话、大湄公河次区域（GMS）经济合作、中亚区域经济合作（CAREC）等现有多边合作机制作用，相关国家加强沟通，让更多国家和地区参与“一带一路”建设。

继续发挥沿线各国区域、次区域相关国际论坛、展会以及博鳌亚洲论坛、中国—东盟博览会、中国—亚欧博览会、欧亚经济论坛、中国国际投资贸易洽谈会，以及中国—南亚博览会、中国—阿拉伯博览会、中国西部国际博览会、中国—俄罗斯博览会、前海合作论坛等平台的建设性作用。支持沿线国家地方、民间挖掘“一带一路”历史文化遗产，联合举办专项投资、贸易、文化交流活动，办好丝绸之路（敦煌）国际文化博览会、丝绸之路国际电影节和图书展。倡议建立“一带一路”国际高峰论坛。

六、中国各地方开放态势推进“一带一路”建设

中国将充分发挥国内各地区比较优势，实行更加积极主动的开放战略，加强东中西互动合作，全面提升开放型经济水平。

西北、东北地区。发挥新疆独特的区位优势和向西开放重要窗口作用，深化与中亚、南亚、西亚等国家交流合作，形成丝绸之路经济带上重要的交通枢纽、商贸物流和文化科教中心，打造丝绸之路经济带核心区。发挥陕西、甘肃综合经济文化和宁夏、青海民族人文优势，打造西安内陆型改革开放新高地，加快兰州、西宁开发开放，推进宁夏内陆开放型经济试验区建设，形成面向中亚、南亚、西亚国家的通道、商贸物流枢纽、重要产业和人文交流基地。发挥内蒙古联通俄蒙的区位优势，完善黑龙江对俄铁路通道和区域铁路网，以及黑龙江、吉林、辽宁与俄远东地区陆海联运合作，推进构建北京—莫斯科欧亚高速运输走廊，建设向北开放的重要窗口。

西南地区。发挥广西与东盟国家陆海相邻的独特优势，加快北部湾经济区和珠江—西江经济带开放发展，构建面向东盟区域的国际通道，打造西南、中南地区开放发展新的战略支点，形成21世纪海上丝绸之路与丝绸之路经济带有机衔接的重要门户。发挥云南区位优势，推进与周边国家的国际运输通道建设，打造大湄公河次区域经济合作新高地，建设成为面向南亚、东南亚的辐射中心。推进西藏与尼泊尔等国家边境贸易和旅游文化合作。

沿海和港澳台地区。利用长三角、珠三角、海峡西岸、环渤海等经济区开放程度高、经济实力强、辐射带动作用大的优势，加快推进中国（上海）自由贸易试验区建设，支持福建建设21世纪海上丝绸之路核心区。充分发挥深圳前海、广州南沙、珠海横琴、福建平潭等开放合作区作用，深化与港澳台合作，打造粤港澳大湾区。推进浙江海洋经济发展示范区、福建海峡蓝色经济试验区和舟山群岛新区建设，加大海南国际旅游岛开发开放力度。加强上海、天津、宁波—舟山、广州、深圳、湛江、汕头、青岛、烟台、大连、福州、厦门、泉州、海口、三亚等沿海城市港口建设，强化上海、广州等国际枢纽机场功能。以扩大开放倒逼深层次改革，创新开放型经济体制机制，加大科技创新力度，形成参与和引领国际合作竞争新优势，成为“一带一路”特别是21世纪海上丝绸之路建设的排头兵和主力军。发挥海外侨胞以及香港、澳门特别行政区独特优势作用，积极参与和助力“一带一路”建设。为台湾地区参与“一带一路”建设作出妥善安排。

内陆地区。利用内陆纵深广阔、人力资源丰富、产业基础较好优势，依托长江中游城市群、成渝城市群、中原城市群、呼包鄂榆城市群、哈长城市群等重点区域，推动区域互动合作和产业集聚发展，打造重庆西部开发开放重要支撑和成都、郑州、武汉、长沙、南昌、合肥等内陆开放型经济高地。加快推动长江中上游地区和俄罗斯伏尔加河沿岸联邦区的合作。建立中欧通道铁路运输、口岸通关协调机制，打造“中欧班列”品牌，建设沟通境内外、连接东中西的运输通道。支持郑州、西安等内陆城市建设航空港、国际陆港，加强内陆口岸与沿海、沿边口岸通关合作，开展跨境贸易电子商务服务试点。优化海关特殊监管区域布局，创新加工贸易模式，深化与沿线国家的产业合作。

七、中国积极行动

一年多来，中国政府积极推动“一带一路”建设，加强与沿线国家的沟通磋商，推动与沿线国家的务实合作，实施了一系列政策措施，努力收获早期成果。

高层引领推动。习近平主席、李克强总理等国家领导人先后出访20多个国家，出席加强互联互通伙伴关系对话会、中阿合作论坛第六届部长级会议，就双边关系和地区发展问题，多次与有关国家元首和政府首脑进行会晤，深入阐释“一带一路”的深刻内涵和积极意义，就共建“一带一路”达成广泛共识。

签署合作框架。与部分国家签署了共建“一带一路”合作备忘录，与一些毗邻国家签署了地区合作和边境合作的备忘录以及经贸合作中长期发展规划，研究编制了与一些毗邻国家的地区合作规划纲要。

推动项目建设。加强与沿线有关国家的沟通磋商，在基础设施互联互通、产业投资、资源开发、经贸合作、金融合作、人文交流、生态保护、海上合作等领域，推进了一批条件成熟的重点合作项目。

完善政策措施。中国政府统筹国内各种资源，强化政策支持。推动亚洲基础设施投资银行筹建，发起设立丝路基金，强化中国—欧亚经济合作基金投资功能。推动银行卡清算机构开展跨境清算业务和支付机构开展跨境支付业务。积极推进投资贸易便利化，推进区域通关一体化改革。

发挥平台作用。各地成功举办了一系列以“一带一路”为主题的国际峰会、论坛、研讨会、博览会，对增进理解、凝聚共识、深化合作发挥了重要作用。

八、共创美好未来

共建“一带一路”是中国的倡议，也是中国与沿线国家的共同愿望。站在新的起点上，中国愿与沿线国家一道，以共建“一带一路”为契机，平等协商，兼顾各方利益，反映各方诉求，携手推动更大范围、更高水平、更深层次的大开放、大交流、大融合。“一带一路”建设是开放的、包容的，欢迎世界各国和国际、地区组织积极参与。

共建“一带一路”的途径是以目标协调、政策沟通为主，不刻意追求一致性，可高度灵活，富有弹性，是多元开放的合作进程。中国愿与沿线国家一道，不断充实完善“一带一路”的合作内容和方式，共同制定时间表、路线图，积极对接沿线国家发展和区域合作规划。

中国愿与沿线国家一道，在既有双多边和区域次区域合作机制框架下，通过合作研究、论坛展会、人员培训、交流访问等多种形式，促进沿线国家对共建“一带一路”内涵、目标、任务等方面的进一步理解和认同。

中国愿与沿线国家一道，稳步推进示范项目建设，共同确定一批能够照顾双多边利益的项目，对各方认可、条件成熟的项目抓紧启动实施，争取早日开花结果。

“一带一路”是一条互尊互信之路，一条合作共赢之路，一条文明互鉴之路。只要沿线各国和衷共济、相向而行，就一定能够谱写建设丝绸之路经济带和21世纪海上丝绸之路的新篇章，让沿线各国人民共享“一带一路”共建成果。

Vision And Actions On Jointly Building Silk Road Economic Belt And 21st Century Maritime Silk Road

Issued by the National Development and Reform Commission, Ministry of Foreign Affairs, and Ministry of Commerce of the People's Republic of China, with State Council authorization

March 2015

First Edition 2015

Preface

More than two millennia ago the diligent and courageous people of Eurasia explored and opened up several routes of trade and cultural exchanges that linked the major civilizations of Asia, Europe and Africa, collectively called the Silk Road by later generations. For thousands of years, the Silk Road Spirit- "peace and cooperation, openness and inclusiveness, mutual learning and mutual benefit" -has been passed from generation to generation, promoted the progress of human civilization, and contributed greatly to the prosperity and development of the countries along the Silk Road. Symbolizing communication and cooperation between the East and the West, the Silk Road Spirit is a historic and cultural heritage shared by all countries around the world.

In the 21st century, a new era marked by the theme of peace, development, cooperation and mutual benefit, it is all the more important for us to carry on the Silk Road Spirit in face of the weak recovery of the global economy, and complex international and regional situations.

When Chinese President Xi Jinping visited Central Asia and Southeast Asia in September and October of 2013, he raised the initiative of jointly building the Silk Road Economic Belt and the 21st Century Maritime Silk Road (hereinafter referred to as the Belt and Road), which have attracted close attention from all over the world. At the China-ASEAN Expo in 2013, Chinese Premier Li Keqiang emphasized the need to build the Maritime Silk Road oriented towards ASEAN, and to create strategic propellers for hinterland development. Accelerating the building of the Belt and Road can help promote the economic prosperity of the countries along the Belt and Road and regional economic cooperation, strengthen exchanges and mutual learning between different civilizations, and promote world peace and development. It is a great undertaking that will benefit people around the world.

The Belt and Road Initiative is a systematic project, which should be jointly built through consultation to meet the interests of all, and efforts should be made to integrate the development strategies of the countries along the Belt and Road. The Chinese government has drafted and published the Vision and Actions on Jointly Building Silk Road Economic Belt and 21st Century Maritime Silk Road to promote the implementation of the Initiative, instill vigor and vitality into the ancient Silk Road, connect Asian, European and African countries more closely and promote mutually beneficial cooperation to a new high and in new forms.

I. Background

Complex and profound changes are taking place in the world. The underlying impact of the international financial crisis keeps emerging; the world economy is recovering slowly, and global development is uneven; the international trade and investment landscape and rules for multilateral trade and investment are undergoing major adjustments; and countries still face big challenges to their development.

The initiative to jointly build the Belt and Road, embracing the trend towards a multipolar world, economic globalization, cultural diversity and greater IT application, is designed to uphold the global free trade regime and the open world economy in the spirit of open regional cooperation. It is aimed at promoting orderly and free flow of economic factors, highly efficient allocation of resources and deep integration of markets; encouraging the countries along the Belt and Road to achieve economic policy coordination and carry out broader and more in-depth regional cooperation of higher standards; and jointly creating an open, inclusive and balanced regional economic cooperation architecture that benefits all. Jointly building the Belt and Road is in the interests of the world community. Reflecting the common ideals and pursuit of human societies, it is a positive endeavor to seek new models of international cooperation and global governance, and will inject new positive energy into world peace and development.

The Belt and Road Initiative aims to promote the connectivity of Asian, European and African continents and their adjacent seas, establish and strengthen partnerships among the countries along the Belt and Road, set up all-dimensional, multi-tiered and composite connectivity networks, and realize diversified, independent, balanced and sustainable development in these countries. The connectivity projects of the Initiative will help align and coordinate the development strategies of the countries along the Belt and Road, tap market potential in this region, promote investment and consumption, create demands and job opportunities, enhance people-to-people and cultural exchanges, and mutual learning among the peoples of the relevant countries, and enable them to understand, trust and respect each other and live in harmony, peace and prosperity.

China's economy is closely connected with the world economy. China will stay committed to the basic policy of opening-up, build a new pattern of all-round opening-up, and integrate itself deeper into the world economic system. The Initiative will enable China to further expand and deepen its opening-up, and to strengthen its mutually beneficial cooperation with countries in Asia, Europe and Africa and the rest of the world. China is committed to shouldering more responsibilities and obligations within its capabilities, and making greater contributions to the peace and development of mankind.

Ⅱ. Principles

The Belt and Road Initiative is in line with the purposes and principles of the UN Charter. It upholds the Five Principles of Peaceful Coexistence: mutual respect for each other's sovereignty and territorial integrity, mutual non-aggression, mutual non-interference in each other's internal affairs, equality and mutual benefit, and peaceful coexistence.

The Initiative is open for cooperation. It covers, but is not limited to, the area of the ancient Silk Road. It is open to all countries, and international and regional organizations for engagement, so that the results of the concerted efforts will benefit wider areas.

The Initiative is harmonious and inclusive. It advocates tolerance among civilizations, respects the paths and modes of development chosen by different countries, and supports dialogues among different civilizations on the principles of seeking common ground while shelving differences and drawing on each other's strengths, so that all countries can coexist in peace for common prosperity.

The Initiative follows market operation. It will abide by market rules and international norms, give play to the decisive role of the market in resource allocation and the primary role of enterprises, and let the governments perform their due functions.

The Initiative seeks mutual benefit. It accommodates the interests and concerns of all parties involved, and seeks a conjunction of interests and the "biggest common denominator" for cooperation so as to give full play to the wisdom and creativity, strengths and potentials of all parties.

Ⅲ. Framework

The Belt and Road Initiative is a way for win-win cooperation that promotes common development and prosperity and a road towards peace and friendship by enhancing mutual understanding and trust, and strengthening all-round exchanges. The Chinese government advocates peace and cooperation, openness and inclusiveness, mutual learning and mutual benefit. It promotes practical cooperation in all fields, and works to build a community of shared interests, destiny and responsibility featuring mutual political trust, economic integration and cultural inclusiveness.

The Belt and Road run through the continents of Asia, Europe and Africa, connecting the vibrant East Asia economic circle at one end and developed European economic circle at the other, and encompassing countries with huge potential for economic development. The Silk Road Economic Belt focuses on bringing together China, Central Asia, Russia and Europe (the Baltic); linking China with the Persian Gulf and the Mediterranean Sea through Central Asia and West Asia; and connecting China with Southeast Asia, South Asia and the Indian Ocean. The 21st Century Maritime Silk Road is designed to go from China's coast to Europe through the South China Sea and the Indian Ocean in one route, and from China's coast through the South China Sea to the South Pacific in the other.

On land, the Initiative will focus on jointly building a new Eurasian Land Bridge and developing China-Mongolia-Russia, China-Central Asia-West Asia and China-Indochina Peninsula economic corridors by taking advantage of international transport routes, relying on core cities along the Belt and Road and using key economic industrial parks as cooperation platforms. At sea, the Initiative will focus on jointly building smooth, secure and efficient transport routes connecting major sea ports along the Belt and Road. The China-Pakistan Economic Corridor and the Bangladesh-China-India-Myanmar Economic Corridor are closely related to the Belt and Road Initiative, and therefore require closer cooperation and greater progress.

The Initiative is an ambitious economic vision of the opening-up of and cooperation among the countries along the Belt and Road. Countries should work in concert and move towards the objectives of mutual benefit and common security. To be specific, they need to improve the region's infrastructure, and put in place a secure and efficient network of land, sea and air passages, lifting their connectivity to a higher level; further enhance trade and investment facilitation, establish a network of free trade areas that meet high standards, maintain closer economic ties, and deepen political trust; enhance cultural exchanges; encourage different civilizations to learn from each other and flourish together; and promote mutual understanding, peace and friendship among people of all countries.

Ⅳ. Cooperation Priorities

Countries along the Belt and Road have their own resource advantages and their economies are mutually complementary. Therefore, there is a great potential and space for cooperation. They should promote policy coordination, facilities connectivity, unimpeded trade, financial integration and people-to-people bonds as their five major goals, and strengthen cooperation in the following key areas:

Policy coordination

Enhancing policy coordination is an important guarantee for implementing the Initiative. We should promote intergovernmental cooperation, build a multi-level intergovernmental macro policy exchange and communication mechanism, expand shared interests, enhance mutual political trust, and reach new cooperation consensus. Countries along the Belt and Road may fully coordinate their economic development strategies and policies, work out plans and measures for regional cooperation, negotiate to solve cooperation-related issues, and jointly provide policy support for the implementation of practical cooperation and large-scale projects.

Facilities connectivity

Facilities connectivity is a priority area for implementing the Initiative. On the basis of respecting each other's sovereignty and security concerns, countries along the Belt and Road should improve the connectivity of their infrastructure construction plans and technical standard systems, jointly push forward the construction of international trunk passageways, and form an infrastructure network connecting all subregions in Asia, and between Asia, Europe and Africa step by step. At the same time, efforts should be made to promote green and low-carbon infrastructure construction and operation management, taking into full account the impact of climate change on the construction.

With regard to transport infrastructure construction, we should focus on the key passageways, junctions and projects, and give priority to linking up unconnected road sections, removing transport bottlenecks, advancing road safety facilities and traffic management facilities and equipment, and improving road network connectivity. We should build a unified coordination mechanism for whole-course transportation, increase connectivity of customs clearance, reloading and multimodal transport between countries, and gradually formulate compatible and standard transport rules, so as to realize international transport facilitation. We should push forward port infrastructure construction, build smooth land-water transportation channels, and advance port cooperation; increase sea routes and the number of voyages, and enhance information technology cooperation in maritime logistics. We should expand and build platforms and mechanisms for comprehensive civil aviation cooperation, and quicken our pace in improving aviation infrastructure.

We should promote cooperation in the connectivity of energy infrastructure, work in concert to ensure the security of oil and gas pipelines and other transport routes, build cross-border power supply networks and power-transmission routes, and cooperate in regional power grid upgrading and transformation.

We should jointly advance the construction of cross-border optical cables and other communications trunk line networks, improve international communications connectivity, and create an Information Silk Road. We should build bilateral cross-border optical cable networks at a quicker pace, plan transcontinental submarine optical cable projects, and improve spatial (satellite) information passageways to expand information exchanges and cooperation.

Unimpeded trade

Investment and trade cooperation is a major task in building the Belt and Road. We should strive to improve investment and trade facilitation, and remove investment and trade barriers for the creation of a sound business environment within the region and in all related countries. We will discuss with countries and regions along the Belt and Road on opening free trade areas so as to unleash the potential for expanded cooperation.

Countries along the Belt and Road should enhance customs cooperation such as information exchange, mutual recognition of regulations, and mutual assistance in law enforcement; improve bilateral and multilateral cooperation in the fields of inspection and quarantine, certification and accreditation, standard measurement, and statistical information; and work to ensure that the WTO Trade Facilitation Agreement takes effect and is implemented. We should improve the customs clearance facilities of border ports, establish a "single-window" in border ports, reduce customs clearance costs, and improve customs clearance capability. We should increase cooperation in supply chain safety and convenience, improve the coordination of cross-border supervision procedures, promote online checking of inspection and quarantine certificates, and facilitate mutual recognition of Authorized Economic Operators. We should lower non-tariff barriers, jointly improve the transparency of technical trade measures, and enhance trade liberalization and facilitation.

We should expand trading areas, improve trade structure, explore new growth areas of trade, and promote trade balance. We should make innovations in our forms of trade, and develop cross-border e-commerce and other modern business models. A service trade support system should be set up to consolidate and expand conventional trade, and efforts to develop modern service trade should be strengthened. We should integrate investment and trade, and promote trade through investment.

We should speed up investment facilitation, eliminate investment barriers, and push forward negotiations on bilateral investment protection agreements and double taxation avoidance agreements to protect the lawful rights and interests of investors.

We should expand mutual investment areas, deepen cooperation in agriculture, forestry, animal husbandry and fisheries, agricultural machinery manufacturing and farm produce processing, and promote cooperation in marine-product farming, deep-sea fishing, aquatic product processing, seawater desalination, marine biopharmacy, ocean engineering technology, environmental protection industries, marine tourism and other fields. We should increase cooperation in the exploration and development of coal, oil, gas, metal minerals and other conventional energy sources; advance cooperation in hydropower, nuclear power, wind power, solar power and other clean, renewable energy sources; and promote cooperation in the processing and conversion of energy and resources at or near places where they are exploited, so as to create an integrated industrial chain of energy and resource cooperation. We should enhance cooperation in deep-processing technology, equipment and engineering services in the fields of energy and resources.

We should push forward cooperation in emerging industries. In accordance with the principles of mutual complementarity and mutual benefit, we should promote in-depth cooperation with other countries along the Belt and Road in new-generation information technology, biotechnology, new energy technology, new materials and other emerging industries, and establish entrepreneurial and investment cooperation mechanisms.

We should improve the division of labor and distribution of industrial chains by encouraging the entire industrial chain and related industries to develop in concert; establish R&D, production and marketing systems; and improve industrial supporting capacity and the overall competitiveness of regional industries. We should increase the openness of our service industry to each other to accelerate the development of regional service industries. We should explore a new mode of investment cooperation, working together to build all forms of industrial parks such as overseas economic and trade cooperation zones and cross-border economic cooperation zones, and promote industrial cluster development. We should promote ecological progress in conducting investment and trade, increase cooperation in conserving eco-environment, protecting biodiversity, and tackling climate change, and join hands to make the Silk Road an environment-friendly one.

We welcome companies from all countries to invest in China, and encourage Chinese enterprises to participate in infrastructure construction in other countries along the Belt and Road, and make industrial investments there. We support localized operation and management of Chinese companies to boost the local economy, increase local employment, improve local livelihood, and take social responsibilities in protecting local biodiversity and eco-environment.

Financial integration

Financial integration is an important underpinning for implementing the Belt and Road Initiative. We should deepen financial cooperation, and make more efforts in building a currency stability system, investment and financing system and credit information system in Asia. We should expand the scope and scale of bilateral currency swap and settlement with other countries along the Belt and Road, open and develop the bond market in Asia, make joint efforts to establish the Asian Infrastructure Investment Bank and BRICS New Development Bank, conduct negotiation among related parties on establishing Shanghai Cooperation Organization (SCO) financing institution, and set up and put into operation the Silk Road Fund as early as possible. We should strengthen practical cooperation of China-ASEAN Interbank

Association and SCO Interbank Association, and carry out multilateral financial cooperation in the form of syndicated loans and bank credit. We will support the efforts of governments of the countries along the Belt and Road and their companies and financial institutions with good credit-rating to issue Renminbi bonds in China. Qualified Chinese financial institutions and companies are encouraged to issue bonds in both Renminbi and foreign currencies outside China, and use the funds thus collected in countries along the Belt and Road.

We should strengthen financial regulation cooperation, encourage the signing of MOUs on cooperation in bilateral financial regulation, and establish an efficient regulation coordination mechanism in the region. We should improve the system of risk response and crisis management, build a regional financial risk early-warning system, and create an exchange and cooperation mechanism of addressing cross-border risks and crisis. We should increase cross-border exchange and cooperation between credit investigation regulators, credit investigation institutions and credit rating institutions. We should give full play to the role of the Silk Road Fund and that of sovereign wealth funds of countries along the Belt and Road, and encourage commercial equity investment funds and private funds to participate in the construction of key projects of the Initiative.

People-to-people bond

People-to-people bond provides the public support for implementing the Initiative. We should carry forward the spirit of friendly cooperation of the Silk Road by promoting extensive cultural and academic exchanges, personnel exchanges and cooperation, media cooperation, youth and women exchanges and volunteer services, so as to win public support for deepening bilateral and multilateral cooperation.

We should send more students to each other's countries, and promote cooperation in jointly running schools. China provides 10,000 government scholarships to the countries along the Belt and Road every year. We should hold culture years, arts festivals, film festivals, TV weeks and book fairs in each other's countries;

cooperate on the production and translation of fine films, radio and TV programs; and jointly apply for and protect World Cultural Heritage sites. We should also increase personnel exchange and cooperation between countries along the Belt and Road.

We should enhance cooperation in and expand the scale of tourism; hold tourism promotion weeks and publicity months in each other's countries; jointly create competitive international tourist routes and products with Silk Road features; and make it more convenient to apply for tourist visa in countries along the Belt and Road. We should push forward cooperation on the 21st Century Maritime Silk Road cruise tourism program. We should carry out sports exchanges and support countries along the Belt and Road in their bid for hosting major international sports events.

We should strengthen cooperation with neighboring countries on epidemic information sharing, the exchange of prevention and treatment technologies and the training of medical professionals, and improve our capability to jointly address public health emergencies. We will provide medical assistance and emergency medical aid to relevant countries, and carry out practical cooperation in maternal and child health, disability rehabilitation, and major infectious diseases including AIDS, tuberculosis and malaria. We will also expand cooperation on traditional medicine.

We should increase our cooperation in science and technology, establish joint labs (or research centers), international technology transfer centers and maritime cooperation centers, promote sci-tech personnel exchanges, cooperate in tackling key sci-tech problems, and work together to improve sci-tech innovation capability.

We should integrate existing resources to expand and advance practical cooperation between countries along the Belt and Road on youth employment, entrepreneurship training, vocational skill development, social security management, public administration and management and in other areas of common interest.

We should give full play to the bridging role of communication between political parties and parliaments, and promote friendly exchanges between legislative bodies, major political parties and political organizations of countries along the Belt and

Road. We should carry out exchanges and cooperation among cities, encourage major cities in these countries to become sister cities, focus on promoting practical cooperation, particularly cultural and people-to-people exchanges, and create more lively examples of cooperation. We welcome the think tanks in the countries along the Belt and Road to jointly conduct research and hold forums.

We should increase exchanges and cooperation between non-governmental organizations of countries along the Belt and Road, organize public interest activities concerning education, health care, poverty reduction, biodiversity and ecological protection for the benefit of the general public, and improve the production and living conditions of poverty-stricken areas along the Belt and Road. We should enhance international exchanges and cooperation on culture and media, and leverage the positive role of the Internet and new media tools to foster harmonious and friendly cultural environment and public opinion.

Ⅴ. Cooperation Mechanisms

The world economic integration is accelerating and regional cooperation is on the upswing. China will take full advantage of the existing bilateral and multilateral cooperation mechanisms to push forward the building of the Belt and Road and to promote the development of regional cooperation.

We should strengthen bilateral cooperation, and promote comprehensive development of bilateral relations through multi-level and multi-channel communication and consultation. We should encourage the signing of cooperation Mous or plans, and develop a number of bilateral cooperation pilot projects. We should establish and improve bilateral joint working mechanisms, and draw up implementation plans and roadmaps for advancing the Belt and Road Initiative. In addition, we should give full play to the existing bilateral mechanisms such as joint committee, mixed committee, coordinating committee, steering committee and management committee to coordinate and promote the implementation of cooperation projects.

We should enhance the role of multilateral cooperation mechanisms, make full use of existing mechanisms such as the Shanghai Cooperation Organization (SCO), ASEAN Plus China (10 + 1), Asia-Pacific Economic Cooperation (APEC), Asia-Europe Meeting (ASEM), Asia Cooperation Dialogue (ACD), Conference on Interaction and Confidence-Building Measures in Asia (CICA), China-Arab States Cooperation Forum (CASCF), China-Gulf Cooperation Council Strategic Dialogue, Greater Mekong Sub-region (GMS) Economic Cooperation, and Central Asia Regional Economic Cooperation (CAREC) to strengthen communication with relevant countries, and attract more countries and regions to participate in the Belt and Road Initiative.

We should continue to encourage the constructive role of the international forums and exhibitions at regional and sub-regional levels hosted by countries along the Belt and Road, as well as such platforms as Boao Forum for Asia, China-ASEAN Expo, China-Eurasia Expo, Euro-Asia Economic Forum, China International Fair for Investment and Trade, China-South Asia Expo, China-Arab States Expo, Western China International Fair, China-Russia Expo, and Qianhai Cooperation Forum. We should support the local authorities and general public of countries along the Belt and Road to explore the historical and cultural heritage of the Belt and Road, jointly hold investment, trade and cultural exchange activities, and ensure the success of the Silk Road (Dunhuang) International Culture Expo, Silk Road International Film Festival and Silk Road International Book Fair. We propose to set up an international summit forum on the Belt and Road Initiative.

Ⅵ. China's Regions in Pursuing Opening-Up

In advancing the Belt and Road Initiative, China will fully leverage the comparative advantages of its various regions, adopt a proactive strategy of further opening-up, strengthen interaction and cooperation among the eastern, western and central regions, and comprehensively improve the openness of the Chinese economy.

Northwestern and northeastern regions. We should make good use of Xinjiang's geographic advantages and its role as a window of westward opening-up to deepen communication and cooperation with Central, South and West Asian countries, make it a key transportation, trade, logistics, culture, science and education center, and a core area on the Silk Road Economic Belt. We should give full scope to the economic and cultural strengths of Shaanxi and Gansu provinces and the ethnic and cultural advantages of the Ningxia Hui Autonomous Region and Qinghai Province, build Xi'an into a new focus of reform and opening-up in China's interior, speed up the development and opening-up of cities such as Lanzhou and Xining, and advance the building of the Ningxia Inland Opening-up Pilot Economic Zone with the goal of creating strategic channels, trade and logistics hubs and key bases for industrial and cultural exchanges opening to Central, South and West Asian countries. We should give full play to Inner Mongolia's proximity to Mongolia and Russia, improve the railway links connecting Heilongjiang Province with Russia and the regional railway network, strengthen cooperation between China's Heilongjiang, Jilin and Liaoning provinces and Russia's Far East region on sea-land multi-modal transport, and advance the construction of an Eurasian high-speed transport corridor linking Beijing and Moscow with the goal of building key windows opening to the north.

Southwestern region. We should give full play to the unique advantage of Guangxi Zhuang Autonomous Region as a neighbor of ASEAN countries, speed up the opening-up and development of the Beibu Gulf Economic Zone and the Pearl River-Xijiang Economic Zone, build an international corridor opening to the ASEAN region, create new strategic anchors for the opening-up and development of the southwest and mid-south regions of China, and form an important gateway connecting the Silk Road Economic Belt and the 21st Century Maritime Silk Road. We should make good use of the geographic advantage of Yunnan Province, advance the construction of an international transport corridor connecting China with neighboring countries, develop a new highlight of economic cooperation in the Greater Mekong Sub-region, and make the region a pivot of China's opening-up to South and Southeast Asia. We should promote the border trade and tourism and culture cooperation between Tibet Autonomous Region and neighboring countries such as Nepal.

Coastal regions, and Hong Kong, Macao and Taiwan. We should leverage the strengths of the Yangtze River Delta, Pearl River Delta, west coast of the Taiwan Straits, Bohai Rim, and other areas with economic zones boasting a high level of openness, robust economic strengths and strong catalytic role, speed up the development of the China (Shanghai) Pilot Free Trade Zone, and support Fujian Province in becoming a core area of the 21st Century Maritime Silk Road. We should give full scope to the role of Qianhai (Shenzhen), Nansha (Guangzhou), Hengqin (Zhuhai) and Pingtan (Fujian) in opening-up and cooperation, deepen their cooperation with Hong Kong, Macao and Taiwan, and help to build the Guangdong-Hong Kong-Macao Big Bay Area. We should promote the development of the Zhejiang Marine Economy Development Demonstration Zone, Fujian Marine Economic Pilot Zone and Zhoushan Archipelago New Area, and further open Hainan Province as an international tourism island. We should strengthen the port construction of coastal cities such as Shanghai, Tianjin, Ningbo-Zhoushan, Guangzhou, Shenzhen, Zhanjiang, Shantou, Qingdao, Yantai, Dalian, Fuzhou, Xiamen, Quanzhou, Haikou and Sanya, and strengthen the functions of international hub airports such as Shanghai and Guangzhou. We should use opening-up to motivate these areas to carry out deeper reform, create new systems and mechanisms of open economy, step up scientific and technological innovation, develop new advantages for participating in and leading international cooperation and competition, and become the pace-setter and main force in the Belt and Road Initiative, particularly the building of the 21st Century Maritime Silk Road. We should leverage the unique role of overseas Chinese and the Hong Kong and Macao Special Administrative Regions, and encourage them to participate in and contribute to the Belt and Road Initiative. We should also make proper arrangements for the Taiwan region to be part of this effort.

Inland regions. We should make use of the advantages of inland regions, including a vast landmass, rich human resources and a strong industrial foundation, focus on such key regions as the city clusters along the middle reaches of the Yangtze River, around Chengdu and Chongqing, in central Henan Province, around Hohhot, Baotou, Erdos and Yulin, and around Harbin and Changchun to propel regional interaction and cooperation and industrial concentration. We should build Chongqing into an important pivot for developing and opening up the western region, and make Chengdu, Zhengzhou, Wuhan, Changsha, Nanchang and Hefei leading areas of opening-up in the inland regions. We should accelerate cooperation between regions on the upper and middle reaches of the Yangtze River and their counterparts along Russia's Volga River. We should set up coordination mechanisms in terms of railway transport and port customs clearance for the China-Europe corridor, cultivate the brand of "China-Europe freight trains," and construct a cross-border transport corridor connecting the eastern, central and western regions. We should support inland cities such as Zhengzhou and Xi'an in building airports and international land ports, strengthen customs clearance cooperation between inland ports and ports in the coastal and border regions, and launch pilot e-commerce services for cross-border trade. We should optimize the layout of special customs oversight areas, develop new models of processing trade, and deepen industrial cooperation with countries along the Belt and Road.

Ⅶ. China in Action

For more than a year, the Chinese government has been actively promoting the building of the Belt and Road, enhancing communication and consultation and advancing practical cooperation with countries along the Belt and Road, and introduced a series of policies and measures for early outcomes.

High-level guidance and facilitation. President Xi Jinping and Premier Li Keqiang have visited over 20 countries, attended the Dialogue on Strengthening Connectivity Partnership and the sixth ministerial conference of the China-Arab

States Cooperation Forum, and met with leaders of relevant countries to discuss bilateral relations and regional development issues. They have used these opportunities to explain the rich contents and positive implications of the Belt and Road Initiative, and their efforts have helped bring about a broad consensus on the Belt and Road Initiative.

Signing cooperation framework. China has signed Mous of cooperation on the joint development of the Belt and Road with some countries, and on regional cooperation and border cooperation and mid-and long-term development plans for economic and trade cooperation with some neighboring countries. It has proposed outlines of regional cooperation plans with some adjacent countries.

Promoting project cooperation. China has enhanced communication and consultation with countries along the Belt and Road, and promoted a number of key cooperation projects in the fields of infrastructure connectivity, industrial investment, resource development, economic and trade cooperation, financial cooperation, cultural exchanges, ecological protection and maritime cooperation where the conditions are right.

Improving policies and measures. The Chinese government will integrate its domestic resources to provide stronger policy support for the Initiative. It will facilitate the establishment of the Asian Infrastructure Investment Bank. China has proposed the Silk Road Fund, and the investment function of the China-Eurasia Economic Cooperation Fund will be reinforced. We will encourage bank card clearing institutions to conduct cross-border clearing operations, and payment institutions to conduct cross-border payment business. We will actively promote investment and trade facilitation, and accelerate the reform of integrated regional customs clearance.

Boosting the role of cooperation platforms. A number of international summits, forums, seminars and expos on the theme of the Belt and Road Initiative have been held, which have played an important role in increasing mutual understanding, reaching consensus and deepening cooperation.

Ⅷ. Embracing a Brighter Future Together

Though proposed by China, the Belt and Road Initiative is a common aspiration of all countries along their routes. China is ready to conduct equal-footed consultation with all countries along the Belt and Road to seize the opportunity provided by the Initiative, promote opening-up, communication and integration among countries in a larger scope, with higher standards and at deeper levels, while giving consideration to the interests and aspirations of all parties. The development of the Belt and Road is open and inclusive, and we welcome the active participation of all countries and international and regional organizations in this Initiative.

The development of the Belt and Road should mainly be conducted through policy communication and objectives coordination. It is a pluralistic and open process of cooperation which can be highly flexible, and does not seek conformity. China will join other countries along the Belt and Road to substantiate and improve the content and mode of the Belt and Road cooperation, work out relevant timetables and roadmaps, and align national development programs and regional cooperation plans.

China will work with countries along the Belt and Road to carry out joint research, forums and fairs, personnel training, exchanges and visits under the framework of existing bilateral, multilateral, regional and sub-regional cooperation mechanisms, so that they will gain a better understanding and recognition of the contents, objectives and tasks of the Belt and Road Initiative.

China will work with countries along the Belt and Road to steadily advance demonstration projects, jointly identify programs that accommodate bilateral and multilateral interests, and accelerate the launching of programs that are agreed upon by parties and ready for implementation, so as to ensure early harvest.

The Belt and Road cooperation features mutual respect and trust, mutual benefit and win-win cooperation, and mutual learning between civilizations. As long as all countries along the Belt and Road make concerted efforts to pursue our common goal, there will be bright prospects for the Silk Road Economic Belt and the 21st Century Maritime Silk Road, and the people of countries along the Belt and Road can all benefit from this Initiative.

推进共建“一带一路”教育行动

教育部

2016 年 7 月

推进共建“丝绸之路经济带”和“21 世纪海上丝绸之路”（以下简称“一带一路”），为推动区域教育大开放、大交流、大融合提供了大契机。“一带一路”沿线国家教育加强合作、共同行动，既是共建“一带一路”的重要组成部分，又为共建“一带一路”提供人才支撑。中国愿与沿线国家一道，扩大人文交流，加强人才培养，共同开创教育美好明天。

一、教育使命

教育为国家富强、民族繁荣、人民幸福之本，在共建“一带一路”中具有基础性和先导性作用。教育交流为沿线各国民心相通架设桥梁，人才培养为沿线各国政策沟通、设施联通、贸易畅通、资金融通提供支撑。沿线各国唇齿相依，教育交流源远流长，教育合作前景广阔，大家携手发展教育，合力推进共建“一带一路”，是造福沿线各国人民的伟大事业。

中国将一以贯之地坚持教育对外开放，深度融入世界教育改革发展潮流。推进“一带一路”教育共同繁荣，既是加强与沿线各国教育互利合作的需要，也是推进中国教育改革发展的需要，中国愿意在力所能及的范围内承担

资料来源：《教育部关于印发〈推进共建“一带一路”教育行动〉的通知》，教育部网站，2016 年 7 月 13 日。

更多责任义务，为区域教育大发展做出更大的贡献。

二、合作愿景

沿线各国携起手来，增进理解、扩大开放、加强合作、互学互鉴，谋求共同利益、直面共同命运、勇担共同责任，聚力构建“一带一路”教育共同体，形成平等、包容、互惠、活跃的教育合作态势，促进区域教育发展，全面支撑共建“一带一路”，共同致力于：

推进民心相通。开展更大范围、更高水平、更深层次的人文交流，不断推进沿线各国人民相知相亲。

提供人才支撑。培养大批共建“一带一路”急需人才，支持沿线各国实现政策互通、设施联通、贸易畅通、资金融通。

实现共同发展。推动教育深度合作、互学互鉴，携手促进沿线各国教育发展，全面提升区域教育影响力。

三、合作原则

育人为本，人文先行。加强合作育人，提高区域人口素质，为共建“一带一路”提供人才支撑。坚持人文交流先行，建立区域人文交流机制，搭建民心相通桥梁。

政府引导，民间主体。沿线国家政府加强沟通协调，整合多种资源，引导教育融合发展。发挥学校、企业及其他社会力量的主体作用，活跃教育合作局面，丰富教育交流内涵。

共商共建，开放合作。坚持沿线国家共商、共建、共享，推进各国教育发展规划相互衔接，实现沿线各国教育融通发展、互动发展。

和谐包容，互利共赢。加强不同文明之间的对话，寻求教育发展最佳契合点和教育合作最大公约数，促进沿线各国在教育领域互利互惠。

四、合作重点

沿线各国教育特色鲜明、资源丰富、互补性强、合作空间巨大。中国将以基础性、支撑性、引领性三方面举措为建议框架，开展三方面重点合作，对接沿线各国意愿，互鉴先进教育经验，共享优质教育资源，全面推动各国教育提速发展。

（一）开展教育互联互通合作

加强教育政策沟通。开展“一带一路”教育法律、政策协同研究，构建沿线各国教育政策信息交流通报机制，为沿线各国政府推进教育政策互通提供决策建议，为沿线各国学校和社会力量开展教育合作交流提供政策咨询。积极签署双边、多边和次区域教育合作框架协议，制定沿线各国教育合作交流国际公约，逐步疏通教育合作交流政策性瓶颈，实现学分互认、学位互授联授，协力推进教育共同体建设。

助力教育合作渠道畅通。推进“一带一路”国家间签证便利化，扩大教育领域合作交流，形成往来频繁、合作众多、交流活跃、关系密切的携手发展局面。鼓励有合作基础、相同研究课题和发展目标的学校缔结姊妹关系，逐步深化拓展教育合作交流。举办沿线国家校长论坛，推进学校间开展多层次多领域的务实合作。支持高等学校依托学科优势专业，建立产学研用结合的国际合作联合实验室（研究中心）、国际技术转移中心，共同应对经济发展、资源利用、生态保护等沿线各国面临的重大挑战与机遇。打造“一带一路”学术交流平台，吸引各国专家学者、青年学生开展研究和学术交流。推进“一带一路”优质教育资源共享。

促进沿线国家语言互通。研究构建语言互通协调机制，共同开发语言互通开放课程，逐步将沿线国家语言课程纳入各国学校教育课程体系。拓展政府间语言学习交换项目，联合培养、相互培养高层次语言人才。发挥外国语院校人才培养优势，推进基础教育多语种师资队伍建设和外语教育教学工作。扩大语言学习国家公派留学人员规模，倡导沿线各国与中国院校合作在华开

办本国语言专业。支持更多社会力量助力孔子学院和孔子课堂建设，加强汉语教师和汉语教学志愿者队伍建设，全力满足沿线国家汉语学习需求。

推进沿线国家民心相通。鼓励沿线国家学者开展或合作开展中国课题研究，增进沿线各国对中国发展模式、国家政策、教育文化等各方面的理解。建设国别和区域研究基地，与对象国合作开展经济、政治、教育、文化等领域研究。逐步将理解教育课程、丝路文化遗产保护纳入沿线各国中小学教育课程体系，加强青少年对不同国家文化的理解。加强“丝绸之路”青少年交流，注重利用社会实践和志愿服务、文化体验、体育竞赛、创新创业活动和新媒体社交等途径，增进不同国家青少年对其他国家文化的理解。

推动学历学位认证标准连通。推动落实联合国教科文组织《亚太地区承认高等教育资历公约》，支持教科文组织建立世界范围学历互认机制，实现区域内双边多边学历学位关联互认。呼吁各国完善教育质量保障体系和认证机制，加快推进本国教育资历框架开发，助力各国学习者在不同种类和不同阶段教育之间进行转换，促进终身学习社会建设。共商共建区域性职业教育资历框架，逐步实现就业市场的从业标准一体化。探索建立沿线各国教师专业发展标准，促进教师流动。

（二）开展人才培养培训合作

实施“丝绸之路”留学推进计划。设立“丝绸之路”中国政府奖学金，为沿线各国专项培养行业领军人才和优秀技能人才。全面提升来华留学人才培养质量，把中国打造成为深受沿线各国学子欢迎的留学目的地国。以国家公派留学为引领，推动更多中国学生到沿线国家留学。坚持“出国留学和来华留学并重、公费留学和自费留学并重、扩大规模和提高质量并重、依法管理和完善服务并重、人才培养和发挥作用并重”，完善全链条的留学人员管理服务体系，保障平安留学、健康留学、成功留学。

实施“丝绸之路”合作办学推进计划。有条件的中国高等学校开展境外办学要集中优势学科，选好合作契合点，做好前期论证工作，构建人才培养模式、运行管理模式、服务当地模式、公共关系模式，使学校顺利落地生根、开花结果。发挥政府引领、行业主导作用，促进高等学校、职业院校与行业企业深化产教融合。鼓励中国优质职业教育配合高铁、电信运营等行业企业走出去，探索开展多种形式的境外合作办学，合作设立职业院校、培训中心，合作开发教学资源和项目，开展多层次职业教育和培训，培养当地急需的各类“一带一路”建设者。整合资源，积极推进与沿线各国在青年就业培训等共同关心领域的务实合作。倡议沿线国家之间开展高水平合作办学。

实施“丝绸之路”师资培训推进计划。开展“丝绸之路”教师培训，加强先进教育经验交流，提升区域教育质量。加强“丝绸之路”教师交流，推动沿线各国校长交流访问、教师及管理人员交流研修，推进优质教育模式在沿线各国互学互鉴。大力推进沿线各国优质教学仪器设备、教材课件和整体教学解决方案输出，跟进教师培训工作，促进沿线各国教育资源和教学水平均衡发展。

实施“丝绸之路”人才联合培养推进计划。推进沿线国家间的研修访学活动。鼓励沿线各国高等学校在语言、交通运输、建筑、医学、能源、环境工程、水利工程、生物科学、海洋科学、生态保护、文化遗产保护等沿线国家发展急需的专业领域联合培养学生，推动联盟内或校际教育资源共享。

（三）共建丝路合作机制

加强“丝绸之路”人文交流高层磋商。开展沿线国家双边多边人文交流高层磋商，商定“一带一路”教育合作交流总体布局，协调推动沿线各国建立教育双边多边合作机制、教育质量保障协作机制和跨境教育市场监管协作机制，统筹推进“一带一路”教育共同行动。

充分发挥国际合作平台作用。发挥上海合作组织、东亚峰会、亚太经合组织、亚欧会议、亚洲相互协作与信任措施会议、中阿合作论坛、东南亚教育部长组织、中非合作论坛、中巴经济走廊、孟中印缅经济走廊、中蒙俄经济走廊等现有双边多边合作机制作用，增加教育合作的新内涵。借助联合国教科文组织等国际组织力量，推动沿线各国围绕实现世界教育发展目标形成协作机制。充分利用中国—东盟教育交流周、中日韩大学交流合作促进委员会、中阿大学校长论坛、中非高校“20＋20”合作计划、中日大学校长论坛、中韩大学校长论坛、中俄大学联盟等已有平台，开展务实教育合作交流。

支持在共同区域、有合作基础、具备相同专业背景的学校组建联盟，不断延展教育务实合作平台。

实施“丝绸之路”教育援助计划。发挥教育援助在“一带一路”教育共同行动中的重要作用，逐步加大教育援助力度，重点投资于人、援助于人、惠及于人。发挥教育援助在“南南合作”中的重要作用，加大对沿线国家尤其是最不发达国家的支持力度。统筹利用国家、教育系统和民间资源，为沿线国家培养培训教师、学者和各类技能人才。积极开展优质教学仪器设备、整体教学方案、配套师资培训一体化援助。加强中国教育培训中心和教育援外基地建设。倡议各国建立政府引导、社会参与的多元化经费筹措机制，通过国家资助、社会融资、民间捐赠等渠道，拓宽教育经费来源，做大教育援助格局，实现教育共同发展。

开展“丝路金驼金帆”表彰工作。对于在“一带一路”教育合作交流和区域教育共同发展中做出杰出贡献、产生重要影响的国际人士、团队和组织给予表彰。

五、中国教育行动起来

中国倡导沿线各国建立教育共同体，聚力推进共建“一带一路”，首先需要中国教育领域和社会各界率先垂范、积极行动。

加强协调推动。加强国内各部门各地方的统筹协调工作，有序开展“一带一路”教育合作交流。推动中国教育治理体系完善、相关法律法规修订和教育综合改革，提升中国开展“一带一路”教育行动的质量和水平。教育部与国家发展改革委、外交部、商务部等部门和全国性行业组织紧密配合，围绕共建“一带一路”大局，寻找合作重点、建立运行保障机制，畅通教育国际合作交流渠道，对接沿线各国教育发展战略规划。

地方重点推进。突出地方推进共建“一带一路”的主体性、支撑性和落地性，要求各地发挥区位优势和地方特色，抓紧制定本地教育和经济携手走出去行动计划，紧密对接国家总体布局。有序与沿线国家地方政府建立“友好省州”“姊妹城市”关系，做好做实彼此间人文交流。充分利用地方调配资源优势，积极搭建海内外平台，促进校企优势互补、良性合作、共同发展。多措并举，支持指导本地教育系统与“一带一路”沿线国家广泛开展合作交流，打造教育合作交流区域高地，助力做强本地教育。

各级学校有序前行。各级各类学校秉承“己欲立而立人”的中国传统，有序与沿线各国学校扩大合作交流，整合优质资源走出去，选择优质资源引进来，兼容并包、互学互鉴，共同提升教育国际化水平和服务共建“一带一路”能力。中小学校要广泛建立校际合作交流关系，重点开展师生交流、教师培训和国际理解教育。高等学校、职业院校要立足各自发展战略和本地区参与共建“一带一路”规划，与沿线各国开展形式多样的合作交流，重点做好完善现代大学制度、创新人才培养模式、提升来华留学质量、优化境外合作办学、助推企业成长等各项工作的协同发展。

社会力量顺势而行。开展更大范围、更深层次、更高水平的“一带一路”教育民间合作交流，吸纳更多民间智慧、民间力量、民间方案、民间行动。大力培育和发展我国非营利组织，通过购买服务、市场调配等举措，大力支持社会机构和专业组织投身教育对外开放事业，活跃民间教育国际合作交流。加快推动教学仪器和中医诊疗服务走出去步伐，支持企业和个人按照市场规则依法参与中外合作办学、合作科研、涉外服务等教育对外开放活动。企业要积极与学校合作走出去，联合开展人才培养、科技创新和成果转化，积极服务“一带一路”国家经贸发展。

助力形成早期成果。实施高度灵活、富有弹性的合作机制，优先启动各方认可度高、条件成熟的项目，明确时间节点，争取短期内开花结果。2016年，各省市制定并呈报本地“一带一路”教育行动计划，有序推进教育互联互通、人才培养培训及丝路合作机制建设。2017年，基于三方面重点合作的沿线各国教育共同行动深入开展。未来3年，中国每年面向沿线国家公派留学生2500人；未来5年，建成10个海外科教基地，每年资助1万名沿线国家新生来华学习或研修。

六、共创教育美好明天

独行快，众行远。合作交流是沿线各国共建“一带一路”教育共同体的主要方式。通过教育合作交流，培养高素质人才，推进经济社会发展，提高沿线各国人民生活福祉，是我们共同的愿望。通过教育合作交流，扩大人文往来，筑牢地区和平基础，是我们共同的责任。

中国愿与沿线各国一道，秉持开放合作、互利共赢理念，共同构建多元化教育合作机制，制订时间表和路线图，推动弹性化合作进程，打造示范性合作项目，满足各方发展需要，促进共同发展。

中国教育部倡议沿线各国积极行动起来，加强战略规划对接和政策磋商，探索教育合作交流的机制与模式，增进教育合作交流的广度和深度，追求教育合作交流的质量和效益，互知互信、互帮互助、互学互鉴，携手推动教育发展，促进民心相通，构建“一带一路”教育共同体，共创人类美好生活新篇章。

Education Action Plan for the Belt and Road Initiative

Issued by the Ministry of Education of the People's Republic of China

July 2016

The "Silk Road Economic Belt and 21st Century Maritime Silk Road" initiative (hereinafter "the Belt and Road Initiative") affords immense opportunities for greater openness, further exchanges, and deeper integration in education in the regions and countries along the routes. Increased cooperation and joint action by the Belt and Road countries in education are an important part of what the Belt and Road Initiative aims to achieve, and in turn, can provide the talent needed to make the Initiative a success. China is ready to work with the countries along the routes to expand people-to-people exchanges, strengthen cooperation in the cultivation of talent, and together create a bright future for education in the region.

Ⅰ. Mission of Education

Education is vital to the strength of a country, the prosperity of a nation and the happiness of a people. It has a fundamental and guiding role to play in the Belt

资料来源：*Education Action Plan for the Belt and Road Initiative*，中国一带一路网，2017 年 10 月 12 日。

and Road Initiative. Educational exchange can serve as a bridge to closer people-to-people ties, whereas the cultivation of talent can buttress the efforts of these countries toward policy coordination, connectivity of infrastructure, unimpeded trade, and financial integration along the routes. The countries along the routes share a closely interdependent bond and educational exchange between these countries goes back a long way, therefore the prospect of educational cooperation is wide and bright. The joining of hands in the development of education to build the Belt and Road together is a great endeavor which will benefit all peoples along the routes.

China will consistently stick to its open policy in education and deeply integrate with the global trends in educational reform and development. Promoting a common prosperity of education in the countries along the routes will not only strengthen win-win cooperation with these countries, but also provide strong stimulus to domestic reform and development in education. China is willing to shoulder as many responsibilities and honour as many commitments as possible, and to make a greater contribution to the development of education in the region.

Ⅱ. Vision for Cooperation

The countries along the routes will work together to deepen mutual understanding, expand openness, strengthen cooperation, learn from each other, to pursue common interests, face our shared future, shoulder common responsibilities, and work concertedly to build a Belt and Road educational community. We will strive for equal, inclusive, mutually beneficial, and dynamic cooperation in education. To promote the development of education in the region and produce wide-ranging support for the Belt and Road Initiative, we will work together in an endeavor to:

1. Promote Closer People-to-People Ties. We will broaden, elevate, and deepen people-to-people exchanges and promote ever-stronger understanding and bonds between the peoples along the routes.

2. Cultivate Supporting Talent. We will spare no effort in cultivating the much-

needed talent for the Belt and Road Initiative to support policy coordination, infrastructure connectivity, unimpeded trade, and financial integration among the Belt and Road countries.

3. Achieve Common Development. We will join hands in deepening educational cooperation and promoting mutual learning to boost the development of education in our countries and improve the overall leverage of the region's education.

Ⅲ. Principles for Cooperation

Principle 1: Focusing on Nurturing of the People, Prioritizing People-to-People Exchanges. To cultivate the talent much needed for the Belt and Road Initiative, we should strengthen cooperation in nurturing of the people and improve the essential competencies of the population throughout the region. We should prioritize people-to-people exchanges and put in place such regional mechanisms that aiming at building a bridge for closer people-to-people ties across the region.

Principle 2: Combining Government Guidance with Social Involvement. Governments of the Belt and Road countries should strengthen mutual communication and coordination, combine various resources and steer the integrated development of education. We should give full play to the proactive role of schools, enterprises, and other social actors to promote dynamic cooperation and diverse exchange in education.

Principle 3: Realizing Shared Growth Through Consultation and Collaboration, and Fostering Greater Openness and Cooperation. With a commitment to the principle of achieving shared growth through consultation and collaboration, the Belt and Road countries should create greater links and better coordination between their national development plans for education, fostering integration and interaction in the development of education of all countries along the routes.

Principle 4: Promoting Harmony, Inclusiveness, Mutual Benefit and Win-Win Outcomes. We should scale up efforts for dialogue between different civilizations, seeking the best corresponding points in educational development and the greatest common measure in educational cooperation. Our ultimate objective is to realize mutual benefit and complementation in education among all countries along the routes.

Ⅳ. Priorities for Cooperation

As each country along the routes has its own distinctive features in education and the whole region in general boasts abundant educational resources, there are great potential for complementation, and vast space for cooperation. China proposes a three-pronged framework of ground-laying, support-building, and forward-thinking actions. Within this framework we propose three main areas of cooperation to speed up the development of education in all Belt and Road countries while catering to each country's educational objectives, referencing to each other's experience in educational development, and sharing each other's best educational resources.

Area 1, we aim to carry out cooperation to improve educational interconnectivity which will include the following 5 elements:

(1) Strengthening coordination on education policy. We will carry out joint studies on how to coordinate domestic laws and policies on education among the Belt and Road countries and put in place an information sharing mechanism for education policies. In doing this, we aim to provide advice to the governments of the Belt and Road countries on coordinating our education policies and to offer policy consultation to schools and social actors in the Belt and Road countries on educational cooperation and exchanges. We will redouble efforts to reach bilateral, multilateral, and sub-regional framework agreements for educational cooperation; propose an international instrument on educational cooperation and exchange among the Belt and Road countries; steadily break policy-related bottlenecks in educational cooperation and exchange, seeking to establish more arrangements to have academic credits to be mutually recognized and more dual and joint degrees conferred, with the prospect of working concertedly to build an integrated educational community.

(2) Facilitating smooth channels for educational cooperation. We will promote simplification of visa application procedures for each other's citizens, expand the scope of educational cooperation and exchange, and forge partnerships featuring frequent interactions, abundant cooperation, dynamic exchanges, and deep affinity. We will encourage schools that already enjoy established cooperative relations, or have similar research projects and share common education development goals, to forge sister school partnerships and steadily deepen and broaden their educational cooperation and exchanges. We will hold forums for school principals, rectors and presidents from countries along the routes and promote pragmatic cooperation of our educational institutions at different levels and in different disciplines. We will support higher-education institutions, based on their respective strengths in specific fields, in establishing joint laboratories (or research centers) and international technology transfer centers with their counterparts in the Belt and Road countries, in a bid to work together to respond to the enormous challenges and opportunities faced by our countries in economic development, resource utilization, ecological preservation, and so on. We will establish academic exchange platforms for the Belt and Road countries, paving the way for our experts, researchers, and students to undertake collaborative research and academic exchanges. We will work to promote sharing of quality educational resources among the Belt and Road countries.

(3) Breaking language barriers between the Belt and Road countries. We will explore how to build coordination mechanisms for breaking language barriers to jointly develop open language courses, and gradually incorporate courses on our different languages into the curricula of each Belt and Road country. We will expand inter-governmental language exchange programs and work together to cultivate, and help each other to cultivate, high-level language experts. We will give full play to the strengths of universities focusing on foreign studies and foreign languages in training strong linguists, and promote the development of multilingual teaching staff for elementary and secondary education as well as foreign language education. We will expand the number of students sent overseas with government scholarships for

language training, and encourage institutions from the Belt and Road countries to work in partnership with Chinese institutions to establish programs that teach their own languages in China. We will support the engagement of more social actors in establishing Confucius Institutes and Confucius Classrooms, and scale up efforts to train both full-time Mandarin teachers and volunteer Mandarin teachers to meet the demand from the Belt and Road countries for Mandarin language training.

(4) Fostering closer people-to-people ties. We will encourage researchers from the countries along the routes to carry out independent and joint research on topics related to China, helping the people of their countries to gain a deeper understanding of China's education and culture, as well as its development model and national policies. We will set up research centers for specific countries and areas and work with counterparts from target countries to carry out research in fields such as economics, politics, education, and culture. We will gradually incorporate international understanding education and Silk Road cultural heritage protection into the curricula of countries along the routes in a bid to deepen the understanding of our youth about cultures in the region. For that matter, we will also boost implementation of youth exchange programs along the Silk Road, giving particular attention to activities and instruments such as internships, volunteer service, cultural experiences, sporting events, innovation and business startups, new media and social media.

(5) Promoting articulation of criteria for mutual recognition of academic credentials. We will strive for the implementation of the UNESCO's Asia-Pacific Regional Convention on the Recognition of Qualifications in Higher Education, support UNESCO's efforts to build a global system for mutual recognition of qualifications, and promote mutual recognition of academic credentials in the region, both bilateral and multilateral. We will encourage countries in the region to improve their education quality assurance systems and academic accreditation mechanisms, and to expedite the development of their national qualifications framework. We will help learners from our different countries to transfer in different types of education or at

different stages in their education, and promote the development of life-long learning societies. We will work to plan together and build together a regional qualifications framework for vocational and technical education and steadily unify employment standards across the region. We will explore the establishment of common standards for teachers'professional development among the Belt and Road countries, delivering greater mobility for teachers within the region.

Area 2, we will deepen cooperation on cultivation and training of talent which will include implementation of the following 4 programs:

(1) Silk Road Two-Way Student Exchange Enhancement Program. The Chinese government will set up the Silk Road Scholarship, aimed at training leading talent and technicians for countries along the routes. China will, attending to every aspect, improve the quality of education received by international students in China and work to turn China into a popular destination for students from the Belt and Road countries. Heading up efforts with national scholarships for study abroad, we will encourage more Chinese students to study in the Belt and Road countries. China will place equal importance on sending students overseas and receiving international students, equal importance on funding students to study overseas and encouraging self-sponsored overseas studies, equal importance on increasing the number of international students and improving the quality of education offered to them, equal importance on law-based management of the students and improving services to them, as well as equal importance on cultivating the talent and giving full play to their roles. China will improve its comprehensive management and service system for outbound and inbound students to ensure their safety, health, and success during their time overseas or in China.

(2) Silk Road Co-Operation in Running Educational Institutions and Programs Enhancement Program. Chinese universities in a position to run educational institutions and programs overseas must concentrate on their strong disciplinary fields, choose the right entry points for cooperation, conduct reliable feasibility surveys, and put in place systems and models for education and training, management and

operations, service to the locality, and public relations with the community and the host countries. This will all be with a view to seeing these educational endeavours operate effectively and perform well in the localities. We will encourage higher education institutions as well as vocational and technical colleges to cooperate with industries and business entities to achieve industry-education integration in which governments in the Belt and Road countries should play a guiding role, while actors in the education sector should play a leading role. We will encourage China's top vocational and technical institutions to develop an overseas presence through collaboration with Chinese high-speed railroad and telecommunication companies, to explore different models of cooperation in running educational institutions and programs overseas, including establishment of vocational colleges, technical colleges and training centers, and joint development of educational resources and programs. Such multi-layered cooperation in vocational and technical education and training will help to cultivate the different kinds of talent that are much needed by the Belt and Road countries. Countries along the routes should all mobilize our resources and actively promote practical cooperation with each other in employment training for young people as well as in other areas of common concern. We call for cooperation among the Belt and Road countries in running top-level educational institutions and programs.

(3) Silk Road Teacher Training Enhancement Program. Through the Silk Road teacher training programs, we aim to strengthen exchanges on the best practices to improve the quality of education in the region. We will encourage teacher exchanges, and give impetus to exchange visits for school principal, and refresher courses and study tours for teaching and management staff in the region. Through this, the Belt and Road countries can learn about the region's best education models in education from each other. We will facilitate the exportation of high-quality teaching equipment, courseware, and full-package teaching solutions from countries along the routes and make forward strides in teacher training with the ultimate objective of achieving equitable distribution of educational resources and balanced development of

education among the Belt and Road countries.

(4) *Silk Road Joint Education and Training Enhancement Program.* We will promote academic exchanges and visits among the Belt and Road countries. We earnestly encourage universities in the region to carry out joint education and training programs needed to meet the urgent development demands of the Belt and Road countries, in fields such as languages, transportation, architecture, medical science, energy, environmental engineering, hydraulic engineering, bio-sciences, marine sciences, ecological preservation, and cultural heritage protection. In this, we aim to facilitate the sharing of educational resources among university alliances and between universities with bilateral relations.

Area 3, we will jointly set up concrete mechanisms of cooperation which include the following 4 elements:

(1) Strengthening high-level consultations on people-to-people exchanges. Through high-level consultations on people-to-people exchange between the Belt and Road countries, both bilateral and multilateral, we will collectively formulate a master blueprint for educational cooperation and exchange in the region. We will coordinate and encourage the establishment by the Belt and Road countries of bilateral or multilateral mechanisms to promote educational cooperation, ensure the quality of education, and oversee the functioning of regional cross-border education markets, so as to coordinate and promote joint action for education in the Belt and Road region.

(2) Giving full play to platforms of international cooperation. To explore new space for educational cooperation among the Belt and Road countries, we will give play to existing bilateral and multilateral mechanisms for cooperation in the region, including the Shanghai Cooperation Organization, the East Asia Summit, Asia-Pacific Economic Cooperation, the Asia-Europe Meeting, the Conference on Interaction and Confidence-Building Measures in Asia, the China-Arab States Cooperation Forum, the Southeast Asian Ministers of Education Organization, the Forum on China-Africa Cooperation, the China-Pakistan Economic Corridor, the Bangladesh-

China-India-Myanmar Economic Corridor, and the China-Mongolia-Russia Economic Corridor. By working with international organizations like the UNESCO, we will give impetus to the establishment of coordination mechanisms among the Belt and Road countries to realize global educational development goals. To further pragmatic educational cooperation and exchanges among the Belt and Road countries, we will fully tap into those platforms already in place, including the China-ASEAN Educational Cooperation Week, the China-Japan-Korea Committee for Promoting Exchange and Cooperation among Universities, the China-Arab States University Presidents Forum, the China-Africa Universities 20 + 20 Cooperation Program, the Forum for Chinese and Japanese University Presidents, the Sino-Korean University Presidents' Forum, and the Sino-Russian University Association. We will support educational institutions within the same region with established cooperative relations and similar disciplinary concentration to form alliances and continue to expand and extend platforms for cooperation in education.

(3) Implementing the "Silk Road Education Assistance Program." Education assistance is an important component of our joint action for education in the Belt and Road region. We will gradually scale up educational assistance, focusing on investing in the people, assisting the people, and benefitting the people. We will give play to the important role of education assistance in "South-South Cooperation," increasing the level of support that goes to countries along the routes, particularly the least-developed countries along the routes. By coordinating governmental and non-governmental educational resources as well as those of the education system, we will educate and train teachers, scholars, and technicians with different specializations for countries in the region. We will actively undertake to provide education assistance packages that incorporate high-quality teaching equipment, teaching schemes, and teacher training. More will be done to strengthen China's education and training centers and the 10 education assistance bases supported by the Chinese Ministry of Education. We encourage that each country establishes mechanisms for diversified fundraising with a leading role by the governments and active involvement

of social actors. By combining government funding, private financing, and public donations, we aim to broaden the funding sources for education, enlarge the scope of education assistance, and achieve shared development in education.

(4) Implementing the "Silk Road Golden Camel & Golden Sail Awards Program." Any individual, team, or organization from the international community that has made an outstanding contribution to promoting educational cooperation and exchange between or among the Belt and Road countries, or that has played an important role in achieving more equitable development of education in the region, is eligible to receive these awards.

V. Chinese Education in Action

China calls for an educational community built by countries along the routes to pool our efforts on the Belt and Road Initiative. This first of all requires that China's education sector and actors from different sectors of Chinese society play a proactive and exemplary role.

1. We should strengthen coordination in the implementation process. We will strengthen overall planning and coordination between different ministries, provinces and cities across China to engage in educational cooperation and exchanges with the Belt and Road countries in an orderly manner. China will improve the governing institution of its education, amend relevant laws and regulations, and proceed with comprehensive educational reform, which will improve its ability to implement smoothly and effectively *the Education Action Plan for the Belt and Road Initiative*. The Chinese Ministry of Education will work closely with other entities like the National Development and Reform Commission, the Ministry of Foreign Affairs, the Ministry of Commerce, and national industrial and trade organizations, with a view to the general development of the Belt and Road Initiative, to identify important areas for cooperation and put safeguarding mechanisms for operation in place to ensure successful implementation. This will help to ensure smoother channels for inter-

national cooperation and exchange in education and to link up with the strategic plans of the Belt and Road countries for the development of education.

2. Different provinces and cities are to have their own focus in moving forward. Prominence will be given to the role of provinces and cities as main actors, the mainstay, and the implementers of the Belt and Road Initiative. Provinces and cities across China are required to use the advantages of their respective geographical position and their distinctive local features as they move quickly to formulate local plans to see that education and the economy go global hand in hand. At the same time, they must closely align their own plans with the national master plan. We encourage local governments in China to steadily establish "friendship province" or "sister city" relationships with their counterparts in countries along the routes, and expect them to effectively promote substantive people-to-people exchanges. Provinces and cities should draw on their own strengths to mobilize resources for establishing domestic and overseas platforms, help universities and businesses to complement each other's strengths, and ensure positive cooperation through which development benefits are shared. Provinces and cities should introduce a range of measures to support and guide local education systems to cooperate with their counterparts in the Belt and Road countries. With this, provinces and cities across China can elevate themselves into prominent ranks in educational cooperation and exchange with overseas counterparts, and in turn, boost their own educational development.

3. Educational institutions at different levels of the education system should steadily press ahead. Guided by the old Chinese motto that "He who craves for success empathizes with others and helps them to be successful," Chinese schools and universities should steadily expand cooperation and exchange with their counterparts in the Belt and Road countries. They should take their best educational resources with them as they engage in cooperation and exchange outside China, and select the most valuable educational resources to bring home from other countries, being inclusive and tolerant, and both learning from and teaching others. In this way together we can make our education more internationalized and strengthen our ability to act in

service of the common development of the Belt and Road Initiative. China's elementary and secondary schools should expand inter-school cooperation and exchange with their counterparts in the Belt and Road countries, concentrating on teacher and student exchanges, teacher training and international understanding education. Chinese universities as well as vocational and technical colleges should work on the basis of their own development strategies and local action plans for the implementation of the Belt and Road Initiative, to carry out various forms of cooperation and exchanges with their counterparts in the Belt and Road countries. Stress should be given to the coordinated development of perfecting the modern institution of university, creating new models for cultivating talent, strengthening the quality of education offered to international students in China, optimizing the performance of overseas educational institutions and programs, and facilitating the growth of busine-sses and enterprises.

4. Social actors should be encouraged to be more involved. Wider-ranging, deeper-delving and higher-aspiring nongovernmental educational cooperation and exchange will pave the way for more insights, support, solutions, and action from our societies. China will scale up its efforts to foster and promote the development of Chinese nonprofit organizations. With government purchase of services and market-based resource allocation, we will give great support to social organizations and specialist groups that dedicate themselves to the cause of international cooperation and exchange in education, and create an enabling environment for international cooperation led by social actors. We will step up the pace of efforts to facilitate the exportation of teaching equipment and traditional Chinese medical services. We will support enterprises and individuals, working according to the rules of the market and in accordance with the law, to engage in forms of international cooperation like cooperation with foreign partners in running educational institutions and programs, carrying out joint research projects, and providing services to foreign customers. Chinese enterprises should work closely with Chinese education providers in exploring opportunities for cooperation in talent cultivation such as technological inno-

vation and technology transfer. This will help to serve economic development and trade growth for the Belt and Road countries.

5. We will work together to yield tangible benefits of cooperation as soon as possible. Highly flexible and resilient mechanisms for cooperation are needed on the ground. We will jumpstart those cooperative projects that are feasible and have been agreed on by stakeholders. We will set clear and reasonable timetables for the implementation of such projects to ensure that they yield the expected benefits within a short span of time. In 2016, Chinese provinces and cities shall submit their own education action plans for the Belt and Road Initiative in which concrete efforts will be made to steadily promote educational connectivity and cultivation of talent, and to establish mechanisms for cooperation. In 2017, the common education action plan of the Belt and Road countries for the Initiative, which prioritizes cooperation in the afore-mentioned three areas, will be further implemented. Over the next three years, China will send annually 2,500 government-funded students to countries along the routes. Over the next five years, China will set up 10 science and education bases overseas and sponsor 10,000 students from the Belt and Road countries to pursue degrees or get short-term training in China.

Ⅵ. Working Together to Create a Bright Future for Education

As an African saying puts it, "He who travels alone travels faster, yet he who travels in company travels farther." Cooperation and exchange are the main venue for the Belt and Road countries to build an educational community. It is our common aspiration to promote economic and social development and improve the living standards of our peoples through educational cooperation and exchange and by working together to cultivate highly competent talent. It is our shared responsibility to strengthen the pillars of regional peace through educational cooperation and exchange and thereby expanding people-to-people exchanges.

China is ready to work with the Belt and Road countries to set up a diverse range of mechanisms for educational cooperation, based on principles of openness, cooperation, mutual-benefit, and win-win outcomes. We will make specific timetables and roadmaps for further cooperation, introduce flexible cooperative mechanisms, and setting up exemplary cooperation projects. This will help us to meet the development demands of each Belt and Road country while at the same time to achieve common development.

The Ministry of Education of China proposes that we, the Belt and Road countries, channel our energies and enthusiasm into action, scale up efforts to align our strategic plans and coordinate our policies, explore new mechanisms and models for educational cooperation and exchange, further deepen and broaden educational cooperation and exchange, and ensure the quality and effectiveness of all such initiatives. Based on principles of mutual understanding, mutual trust, mutual assistance and mutual learning, we shall join hands to promote the development of education and closer people-to-people ties. With these efforts, we will build an educational community among the Belt and Road countries together and create a new chapter of beautiful life for all humanity.

二

“一带一路”建筑类大学国际联盟概况

Overview of the Belt and Road Architectural University International Consortium

关于构建“一带一路”建筑类大学国际联盟的倡议书

一、目的

为了积极响应国家发展改革委、外交部、商务部联合发布的《推动共建丝绸之路经济带和21世纪海上丝绸之路的愿景与行动》及国家“一带一路”倡议，切实贯彻落实中央及国家《关于做好新时期教育对外开放工作若干指导意见》，教育部《推进共建“一带一路”教育行动》，北京市《新时期北京教育对外开放工作规划（2016—2020）》和《北京市对接共建“一带一路”教育行动计划实施方案》等文件精神，北京建筑大学建议并发起构建“一带一路”建筑类大学国际联盟（以下简称“联盟”）。该联盟旨在搭建建筑教育信息共享、学术资源共享的交流合作平台，探索跨国培养与跨境流动的人才培养机制，促进联盟高校之间的流动。联盟成员将协同创新，共同开展建筑领域的合作与交流，共同创新联合机制，共同打造政治互信、经济融合、文化包容的利益共同体、命运共同体和责任共同体。

二、组织架构

我校先后与国内外多所高校及产学研研究生培养基地等多家企业共同组建了“一带一路”建筑类大学国际联盟，联盟成员依照联盟章程开展活动。

三、主要职能

（一）加强联盟成员之间全面合作与交流，共同创建落地项目，不断加深感情。

（二）共同设立联盟研究经费，鼓励大学联盟内的教师开展教学及科学研究，共同申请第三方科学研究项目。

（三）鼓励联盟各成员单位学术人员、行政管理人员、专业技术人员在大学联盟内流动，开展讲学、指导学生、举办学术研讨会、管理经验交流等活动。

（四）积极支持、开展各项汉语推广活动。

（五）建立大学联盟内的学生交流机制，制定相应规则，促进学生在大学联盟内的交流学习，实现学生的学分互认及学术成果互认。

（六）对接中国企业“走出去”战略，积极协助联盟企业，开展提供咨询、培训服务。

（七）定期召开联盟成员大会。

四、运行模式

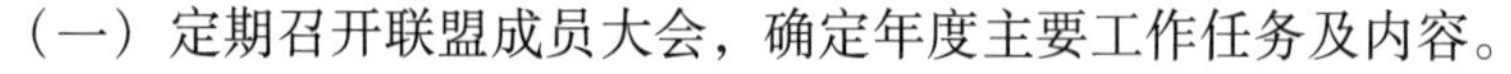

（一）定期召开联盟成员大会，确定年度主要工作任务及内容。

（二）确定年度活动主题，每年举办联盟单位参与的学术、科技、文化、体育、艺术等交流活动。

（三）面对联盟企业，利用双（多）边大学自身的优势，定期为联盟企业提供培训、咨询等相关服务。

（四）定期实现联盟大学成员相关人员的互访。

（五）开展形式多样的学生交流活动，如暑期交流、寒假交流、学期交换，学生社团交流等。

（六）每年举办不同领域、不同主题的学术研讨会。

（七）开展联盟成员内文化研究和区域国别研究。

五、条件保障

（一）积极申请双方政府的相关政策、资金支持。

（二）北京建筑大学预备专项经费支持大学联盟相关活动。

（三）加大联盟成员相关活动的外宣活动。

（四）提供相应的工作人员以及办公条件。

北京建筑大学

2017 年 5 月

Proposal For Establishing Belt and Road Architectural University International Consortium Under The Context of The Belt and Road Initiative

I. Objectives

To actively respond to China's Belt and Road strategy and *Vision And Actions On Jointly Building Silk Road Economic Belt And 21st Century Maritime Silk Road* issued by the National Development and Reform Commission, Ministry of Foreign Affairs, and Ministry of Commerce of the People's Republic of China and to effectively implement *Several Opinions on Successfully Opening Up Education in the New Period* issued by the central government, *Actions on Jointly Promoting Education Along the Belt and Road* issued by the Ministry of Education, *Beijing's Work Plan for Opening Up Education in the New Period* (2016 – 2020), *Beijing's Implementation Plan for Participating in Promoting Education Along the Belt and Road*, the Belt and Road Architectural University International Consortium initiated by Beijing University of Civil Engineering and Architecture (BUCEA) is forged to build a communication and cooperation platform for sharing information and academic resources and to explore a new type of talent training model featuring cross-cultural training and cross-border flow of talents. Member universities work together to innovate, to promote cooperation and exchange in the field of architecture, and to build

a community of shared interests, destiny and responsibility featuring mutual political trust, economic integration and cultural inclusiveness.

Ⅱ. Organizational structure

The Consortium consists of universities, Industry-University-Research graduate students cultivation bases and enterprises. All members conduct activities according to the Charter.

Ⅲ. Major functions

3.1 Enhance exchanges and communication among member universities, jointly create landing projects and promote people-to-people bonds.

3.2 Jointly set up a Consortium Research Fund to stimulate cooperation among faculties in lecture-giving, scientific research and third-party research projects application.

3.3 Encourage the mobility and exchanges of academic, administrative, professional and technical staff within the Consortium and jointly organize academic exchange activities like lectures, student-tutoring and academic seminars.

3.4 Support and carry out activities to promote the Chinese language.

3.5 Establish students exchange mechanisms to promote students mobility among the member universities, and formulate relevant rules to ensure mutual recognition of students' credits and academic achievements.

3.6 In accordance with Chinese enterprises' "Going Abroad" strategy, help Chinese enterprise through offering consultation and training assistance.

3.7 Hold general meeting of Consortium on a regular basis.

Ⅳ. Operation Model

4.1 Hold general meeting of Consortium on a regular basis to confirm each year's job list and specifications.

4.2 Identify each year's theme for a series of academic, scientific, cultural, sports and art activities among member universities.

4.3 Offer training and consultation assistance to member enterprises on a regular basis.

4.4 Promote regular mutual visits among staff of member universities.

4.5 Carry out a variety of student exchange activities including summer programs, winter programs, semester exchange programs and student union exchange programs.

4.6 Organize academic seminars of various domains and themes.

4.7 Carry out cultural studies, and national and regional studies in each country involved.

Ⅴ. Supporting Conditions

5.1 Actively apply for government support and fund from the member countries.

5.2 Each member university should reserve special funds to support the Consortium's relevant activities.

5.3 Strengthen the publicity for the Consortium.

5.4 Provide necessary personnel and working conditions.

Beijing University of Civil Engineering and Architecture

May, 2017

“一带一路”建筑类大学国际联盟章程

第一章　总则

第一条　联盟名称：“一带一路”建筑类大学国际联盟。

第二条　联盟性质：在北京市教委、北京市人民政府外事办公室的支持与指导下，在现有合作的基础上，由北京建筑大学发起成立“一带一路”建筑类大学国际联盟，由国内外多所大学组成。联盟性质为按照“自愿平等、开放共享、合作共赢、创新发展”的原则自发组织的非营利性战略合作组织。

第三条　联盟宗旨：发挥特色优势、推进资源共享、加强协同创新、促进人才培养，提升国际科研合作水平，促进跨文化交流与理解，促进大学间人员、知识、技术等各方面资源的流动，为培养高素质国际化人才，提升国际交流合作的能力与水平，服务经济结构转型升级，促进民心相通、跨国界产学研合作，提升大学内涵实力和国际声誉而不断努力。

第二章　任务

第四条　主要职能

1. 致力于高素质、国际化工程技术人才培养和培育模式的创新实践，为社会培养高水平、实践能力强的工程技术人才，实现高层次工程技术人才教育的新突破。

2. 致力于促进经济结构转型、产业升级以及集成解决社会共性关键技术为目标，以科研项目和技术创新为牵引，创新合作机制，打造跨国界多校对社会的协同创新平台，促进资金、产品、人才和服务的跨国界流动。

3. 致力于推进民心相通、跨文化交流与理解。积极促进大学间跨国界的人员和文化交流，鼓励大学间人员跨国界流动，联合举办各类学术会议、科技竞赛、各类文化、艺术及体育类等活动，积极开展各项汉语推广活动。

第五条　业务范围

1. 秉承开放式办学、合作共赢的思想，充分发挥成员学校的优势，通过资源共享、优势互补、协同创新和强强合作，在人才培养、科学研究、师资队伍建设、校园文化、社会服务等方面开展全方位的合作与交流。

2. 人才培养：开展工程技术高素质国际化人才培养的协同创新，建立建筑领域国际化优秀人才培养实验区，建立具有国际国内先进水平的实训基地，建立学分互认、学生跨国界、跨校区流动机制，建立精品基础课及学科前沿讲座等资源的共享、以及优秀教师跨国界任课等制度，使学生在跨文化环境中接触到更宽泛的学科领域，学习到各领域前沿、先进的知识和技术，掌握跨文化交流与合作的能力和技巧，培养高素质国际化复合型人才。

3. 科学研究：发挥各自特色优势，形成合力，协同创新，联合项目攻关，共同解决国家需求和制约社会发展的重大科学技术问题。促进各校优势特色学科的交叉融合，整合各校科研力量，并与科研机构、企业深度合作，共同承担重大科技攻关项目，集成解决国家支柱产业核心关键技术，提高自主创新能力，更好地促进行业技术进步、推动产业结构调整。

4. 师资队伍建设：开展教学经验、科研成果的交流和互动，召开高水平国际会议，开展高层次学术访问、研究生导师互聘等活动，不断提升教学科研水平，建设高水平的教师队伍。

5. 校园文化：开展大学间学生体育、艺术、科技竞赛以及学生暑期、寒假短期文化体验等文化交流，增进彼此间的友谊，促进学生的全面发展。开展汉语教学和中国文化推广活动，促进文化的交流与理解。

6. 社会服务：积极发挥外专智库作用。为政府和社会组织实现人才、资

本、产品、服务等跨国界的咨询、培训、政策、法律和技术支持等服务。

第三章 成员

第六条 由国内外合作院校共同组建“一带一路”建筑类大学国际联盟，联盟成员依照联盟章程开展活动。

第七条 申请加入联盟的学校需承认并遵守联盟的章程，具有显著的办学特色和突出的学科群优势，经联盟理事会批准后即可成为联盟成员。成员单位根据需要可适当增减。

第八条 成员权利

1. 申请加入自愿，申请退出自由；
2. 联盟内的选举权和被选举权；
3. 对联盟重大事件决策的表决权；
4. 对联盟日常工作的参与、批评、建议和监督权；
5. 参加、承办或协办联盟举办的各项活动；
6. 利用联盟平台获得相关信息资源。

第九条 成员义务

1. 遵守联盟章程，维护联盟权益；
2. 执行联盟决议，完成联盟委托的工作；
3. 积极参加、轮流承办或协办联盟举行的活动；
4. 充分尊重联盟成员间的知识产权。

第十条 加入联盟的程序

1. 大学提出申请，提交秘书处预先审核；
2. 理事会开会讨论并表决通过。

第十一条 退出联盟的程序

1. 有退出意愿的成员，须向联盟提交退出的书面申请；
2. 经秘书处审核，理事会讨论通过。

第四章　组织与运行

第十二条　联盟实行理事会制度，理事会下设秘书处。

第十三条　联盟理事会

1. 理事会是联盟的决策机构。

2. 理事会成员构成：联盟各大学为成员单位，成员单位负责人为理事会成员。

3. 理事会主要职责：制定和修改联盟章程、制定联盟基本规章制度。选举理事会领导机构成员，制定联盟年度工作方案，审议理事会年度工作报告或理事提出的议案，研讨制定联盟合作与发展规划，统筹各联盟成员的资源与共享，协调各成员单位的关系，决定联盟成员的加入与退出。

4. 理事会主席实行轮值制度，每两年轮值一次。轮值主席的主要职责为主持召开理事会议，组织讨论、审定联盟年度计划，牵头协调解决联盟重大问题等。理事会副主席，由轮值主席提名并经联盟理事会讨论和批准产生。负责组织起草分管工作的年度计划，组织实施联盟年度工作方案，向理事会提交联盟发展有关议案，负责协调分管工作的联盟各成员之间关系。

5. 理事会一般每年召开 1 次会议，如遇特殊情况，可由理事提议、召开临时理事会。理事会按民主集中制原则议事，决定重大问题须经半数以上理事同意。

第十四条　秘书处

1. 秘书处是联盟的常设办事机构，办公地点设在联盟主席所在单位。联盟设秘书长 1 名，副秘书长 3 名。秘书长人选由理事会主席推荐。

2. 秘书处主要职责：掌握统筹联盟各单位间的人才培养、科技、学科、师资等资源，建立联盟数据库并进行实时更新；负责联盟对外合作管理；完成理事主席和副主席交办的日常工作；牵头负责联盟年度工作方案的落实；筹备组织理事会议等；创办联盟网站并维护其正常运行；负责联盟的联络、协调工作。

第五章　附则

第十五条　本章程的修订由理事会提出，经理事会会议讨论通过后生效。

第十六条　理事会可按照章程的规定，制定章程细则。章程细则不得与章程的规定相抵触。

第十七条　本章程经理事会表决通过生效。

第十八条　本章程的解释权属理事会。

Charter of the Belt and Road Architectural University International Consortium

Chapter 1 General Rules

Article 1. Name of the Consortium: Belt and Road Architectural University International Consortium (BRAUIC).

Article 2. Nature of the Consortium: Under the support and guidance of Beijing Municipal Education Commission and Foreign Affairs Office of the People's Government of Beijing Municipality, Beijing University of Civil Engineering & Architecture put forward the proposal to build the Belt and Road Architectural University International Consortium on the basis current connections with partner universities from home and abroad. The Consortium is voluntarily organized for non-profit strategic cooperation under the principle of "equality and free will, openness and sharing, win-win cooperation, and innovative development".

Article 3. Purpose of the Consortium: Explore each member's unique advantages, promote resources sharing, enhance coordinated innovation, facilitate cultivation of talent, accelerate international cooperation, cultural exchanges and mutual understanding, facilitate the exchanges of staff, knowledge and technology among different member universities, and make continuous efforts to cultivate high-quality international talents, improve capabilities and standard of international exchange and cooperation, enhance people-to-people bonds, international industry-university-research collaboration and improve the actual strength and international reputation of the member universities.

Chapter 2 Tasks

Article 4. Major functions

4.1 Committed to making breakthroughs in the cultivation of high-quality international engineering and technological talents, cultivating innovation, and delivering qualified engineering and technical personnel with practical ability.

4.2 Committed to facilitating economic structure transformation, industry upgrading, and to offering integrated solution to common and critical technology needs. The Consortium plans to take S&T projects and technological innovation as the focus, and build platforms for universities from different nations to coordinate innovation efforts, so as to stimulate the exchanges of capital, products, personnel and service among different countries.

4.3 Committed to fostering people-to-people bonds, enhancing cross-cultural understanding by actively promoting cultural exchanges. The Consortium plans to encourage staff and students' mobility trans-nationally, to jointly organize various academic conferences, S&T competition, cultural, arts and sports activities, etc. and to carry out activities to promote the Chinese language.

Article 5. Operation scope

5.1 In line with the spirit of open operation and win-win cooperation, the Consortium seeks to give full play to each member's unique advantages. Through resources sharing, complementing each other's strengths, coordinated innovation and strength-to-strength cooperation, the Consortium is committed to carrying out all-rounded cooperation and exchanges in areas like talents cultivation, scientific research, faculty member development, campus culture development and social services.

5.2 Talent cultivation

Innovate training mode so as to cultivate high-level international technological talents. Establish internationally-advanced experimental talents training base. Provide mechanisms for students'credit transfer and students'transnational and cross-campus mobility. Find ways for member universities to share their basic courses and frontier knowledge lectures, encourage excellent teachers to give lectures in different nations and campus. Offer students access to broader disciplines so that they can acquire the most up-to-date knowledge and technology in a cross-cultural environment where they can increase their cross cultural communication capacity and skills.

5.3 Scientific research

Give full play to respective advantages, form synergy, innovate collaboratively, jointly develop research projects and solve major scientific and technological problems restricting the development. Promote integration of advantageous disciplines with distinguished features of the universities, integrate scientific research resources, facilitate deeper cooperation with scientific research institutions and enterprises, jointly work on major S&T projects and solve the key technologies of national pillar industries, improve the ability of independent innovation and better promote the technical progress and structure adjustment of the industries.

5.4 Faculty member development

Develop exchange of teaching experience and scientific research achievements, hold high-level international conferences, carry out high-level academic visits and mutual employment of postgraduate student supervisors, improve the teaching and scientific research level in the development of high-level faculty.

5.5 Campus culture

Carry out sports, arts, scientific and technological competitions and short-term cultural exchanges in summer and winter vacations between universities to enhance friendship and promote the comprehensive development of students. Hold Chinese teaching and culture promotion activities and promote cultural communication and understanding.

5.6 Social services

Take great advantage of think tanks and provide support in consulting, training, policy, law and technology for communication and cooperation among governments and social organizations so as to facilitate the flow of talents, capital, products and services.

Chapter 3 Members

Article 6. The Consortium consists of universities from different countries. All members conduct activities according to the Charter.

Article 7. The universities applying for admission to the Consortium shall accept and abide by the Charter and have distinctive features in education operation and outstanding advantages in disciplines. The universities become Consortium members upon approval of the Consortium council.

Article 8. Member rights

8. 1 Voluntary to join and free to withdraw;

8. 2 Right to vote and to be elected in the Consortium;

8. 3 Right to vote on major events in the Consortium;

8. 4 Right to participate in, criticize, suggest and supervise the routine of the Consortium;

8. 5 Participate in, host or co-organize activities of the Consortium;

8. 6 Obtain relevant information resources via the Consortium platform.

Article 9. Member obligations

9. 1 Abide by the Charter and maintain rights and interests of the Consortium;

9. 2 Carry out the resolutions and complete the work entrusted by the Consortium;

9. 3 Actively participate in, take turns to host or co-organize the events of the Consortium;

9. 4 Fully respect the intellectual property rights among Consortium members.

Article 10. Procedures to join the Consortium

10.1 A prospect member file an application for a preliminary review by the Secretariat;

10.2 Discuss and vote to approve at the council meeting.

Article 11. Procedures to withdraw from the Consortium

11.1 Submit written application of withdrawal to the Consortium;

11.2 Reviewed by the Secretariat and approved by the council after discussion.

Chapter 4 Organization and Operation

Article 12. The Consortium implements the council system with the Secretariat under the Council.

Article 13. Consortium Council

13.1 The Council is the decision-making body of the Consortium.

13.2 Members of the Council: universities in the Consortium are the member units. Some representatives from member universities would be admitted to the Council.

13.3 Main responsibilities: formulate and revise the Charter of the Consortium, make fundamental regulations and rules of the Consortium, elect leadership members of the council, develop annual work plan, review annual report or proposals of the council, discuss and formulate cooperation and development planning of the Consortium, coordinate resource sharing and relationships among Consortium members, and resolve on the joining and withdrawal of Consortium members.

13.4 The Consortium Council implements a rotating presidency for every two years. The main responsibilities of the Chairman are as follows: preside over the council meetings, discuss and examine the annual plan of the Consortium, take the lead in solving the major problems of the Consortium, etc. The Vice Chairmen are nominated by Chairman and should be ratified by the Consortium Council. The main responsibilities of the Vice Chairmen include the following: organize the draft of annual plans, organize the implementation of resolutions of the council, submit development proposals to the council, and coordinate relationship among Consortium members.

13. 5 The Council will generally hold an annual meeting. In case of special circumstances, the council might propose to convene an interim council. Based on principles of democratic centralism, the major issues must be approved by more than half of the council members.

Article 14. The Secretariat

14. 1 The Secretariat is the standing office of the Consortium. The office is located at the Chairman's university. There is one Secretary-General and 3 Deputy Secretaries-General. The candidacy of Secretary-General is recommended by the Chairman of the council.

14. 2 Main responsibilities of the Secretariat: coordinate the resources of talent training, science and technology, disciplines, faculty, etc. , build Consortium database and do real-time update, manage co-operation with foreign countries, complete routine work assigned by the Chairman and Vice Chairmen, implement annual work plan, prepare and organize the council meetings, establish and maintain Consortium website, and take charge of liaison and coordination.

Chapter 5 Supplementary Articles

Article 15. Revision to this Charter shall be proposed by the Council and take effect after discussion and approval at the council meeting.

Article 16. The Council may formulate provisions in accordance with the articles of the Charter. The provisions shall not contravene the articles of the Charter.

Article 17. This Charter take effect after being approved by the council.

Article 18. Interpretation of this Charter belongs to the Council.

"一带一路"建筑类大学国际联盟成员

Members Of The Belt and Road Architectural University International Consortium

洲别 Continent	国家/地区 Country/ Area	序号 S. N.	成员单位 Members
亚洲 Asia	中国 China	1	安徽建筑大学 Anhui Jianzhu University
		2	北京建筑大学 Beijing University of Civil Engineering and Architecture
		3	重庆交通大学 Chongqing Jiaotong University
		4	河北建筑工程学院 Hebei University of Architecture
		5	吉林建筑大学 Jilin Jianzhu University
		6	山东建筑大学 Shandong Jianzhu University
		7	沈阳建筑大学 Shenyang Jianzhu University
		8	天津城建大学 Tianjin Chengjian University
		9	西安建筑科技大学 Xi'an University of Architecture and Technology

续表

洲别 Continent	国家/地区 Country/ Area	序号 S. N.	成员单位 Members
亚洲 Asia	韩国 Republic of Korea	10	大田大学 Daejeon University
	马来西亚 Malaysia	11	马来西亚大学联盟 Consortium of Malaysian University
		12	马来西亚理工大学 University of Technology Malaysia
	印度 India	13	印度科技教育集团 Techno India Group
	以色列 Israel	14	贝扎雷艺术与设计学院 Bezalel Academy of Art and Design
	吉尔吉斯斯坦 Kyrgyzstan	15	奥什工业大学 Osh Technological University
	尼泊尔 Nepal	16	尼泊尔工程学院 Nepal Engineering College
欧洲 Europe	俄罗斯联邦 Russian Federation	17	莫斯科罗蒙诺索夫国立大学数学力学学院 Faculty of Mathematics & Mechanics, Lomonosov Moscow State University
		18	伊尔库茨克国家研究型技术大学 Irkutsk National Research Technical University
		19	喀山联邦大学 Kazan (Volga Region) Federal University
		20	莫斯科国立建筑学院 Moscow Institute of Architecture (State Academy)
		21	莫斯科大学 Moscow State University
		22	国家研究型莫斯科国立建筑大学 National Research Moscow State University of Civil Engineering
		23	俄罗斯建筑土木科学院 Russian Academy of Architecture and Construction Sciences

续表

洲别 Continent	国家/地区 Country/ Area	序号 S. N.	成员单位 Members
欧洲 Europe	俄罗斯联邦 Russian Federation	24	南联邦大学 Southern Federal University
		25	圣彼得堡技术大学 St. Petersburg State Polytechnical University
		26	圣彼得堡国立建筑大学 St. Petersburg State University of Architecture and Civil Engineering
		27	秋明工业大学 Tyumen Industrial University
		28	乌拉尔联邦大学 Ural Federal University
		29	乌拉尔国立建筑艺术学院 Ural State Academy of Architecture and Arts
	波兰 Poland	30	琴希托霍瓦理工大学 Czestochowa University of Technology
		31	西里西亚理工大学 Silesian University of Technology
		32	华沙生态与管理大学 University of Ecology and Management in Warsaw
		33	华沙理工大学 Warsaw University of Technology
	法国 France	34	马恩·拉瓦雷大学 Université de Marne-La-Vallée
		35	英科工程大学联盟 Yncréa Hauts-De-France
	亚美尼亚 Armenia	36	亚美尼亚国立建筑大学 Armenia State University of Architecture and Construction
	保加利亚 Bulgaria	37	保加利亚土木建筑及大地测量大学 University of Architecture, CivilEngineering and Geodesy
	捷克共和国 Czech Republic	38	南波西米亚大学 South Bohemia University
	希腊 Greece	39	塞萨洛尼基亚里士多德大学 Aristotle University of Thessaloniki

续表

洲别 Continent	国家/地区 Country/Area	序号 S. N.	成员单位 Members
欧洲 Europe	挪威 Norway	40	卑尔根建筑学院 Bergen School of Architecture
	意大利 Italy	41	蒙塞拉特基金会 Monserrate Foundation
		42	罗马第三大学 Roma Tre University
	塞尔维亚 Serbia	43	诺维萨德大学 Novi Sad University
	罗马尼亚 Romania	44	斯皮鲁哈列德大学 Spiru Haret University
	土耳其 Turkey	45	伊斯坦布尔服装学院 Istanbul Arel University
	英国 Britain	46	德比大学 University of Derby
		47	东伦敦大学 University of East London
		48	肯特大学 University of Kent
美洲及大洋洲 America and Oceania	美国 the United States	49	夏威夷太平洋大学 Hawai'i Pacific University
		50	里海大学 Lehigh University
		51	夏威夷大学建筑学院 School of Architecture, University of Hawai'i
	澳大利亚 Australia	52	悉尼科技大学 University of Technology, Sydney
	巴西 Brazil	53	圣保罗大学 University of São Paulo

三

“一带一路”建筑类大学国际联盟大会暨校长论坛

Conferences and Presidents' Forums of the Belt and Road Architectural University International Consortium

“一带一路”建筑类大学国际联盟在北京建筑大学成立

2017年10月10日上午，“一带一路”建筑类大学国际联盟成立大会和校长论坛在北京建筑大学大兴校区图书馆建本报告厅隆重举行（图1、图2）。北京市教育委员会副主任葛巨众，来自国内外25所建筑类大学的校长和代表，媒体的参会代表等，共计百余人参会，共同见证“一带一路”建筑类大学国际联盟正式成立，并围绕创新推进“一带一路”倡议下建筑类大学国际交流与合作，以及创新人才培养展开深入研讨交流。

图1　“一带一路”建筑类大学国际联盟成立大会在北京建筑大学召开

图 2　“一带一路”建筑类大学国际联盟在北京建筑大学成立

北京建筑大学党委书记王建中主持联盟成立大会并致欢迎辞（图 3）。他讲到，作为“一带一路”建设的重要内容，教育在共建“一带一路”中具有基础性和先导性作用。“一带一路”重大倡议的提出，为提高沿线国家和地区教育水平、推动各国间人文交流和区域繁荣发展提供了重大机遇。面对“一带一路”建设中基础设施互联互通的重大历史机遇，以及对建筑类人才和科技的迫切需求，成立“一带一路”建筑类大学国际联盟，推动“一带一路”沿线建筑类大学交流合作，创新建筑类人才培养机制，探索跨文化建筑类学科专业建设，推进科技成果转化，服务“一带一路”沿线及欧亚地区的发展建设，全面投身、积极融入这一伟大实践是我们建筑类高校责无旁贷的历史使命，对推动实现“一带一路”教育共同繁荣具有重要意义。“一带一路”建筑类大学国际联盟的成立，标志着沿线国家建筑类大学的合作已经迈出了坚实的第一步，我们将进一步深化务实合作，推动互利共赢，建立更多合作交流机制，为推进世界城市和建筑事业繁荣发展、增进人类和谐宜居福祉做出更大贡献。

图 3　北京建筑大学党委书记王建中

联盟首届轮值主席、北京建筑大学校长张爱林以“把倡议变为行动，把愿景变为现实，创新推进‘一带一路’建筑类大学国际交流合作”为题报告了联盟成立背景、章程及合作意向。（图 4）他谈到，最近两年，北京建筑大学与“一带一路”沿线国家高校开展了广泛合作，发展了一批新的友好伙伴高校，签署校际合作协议，双方进行师生交流、开展科研合作、共同举办国际会议，开启了交流合作的新篇章。2016 年，学校获得北京市教委“一带一路”沿线国家留学生专项奖学金，用于资助“一带一路”沿线国家有志青年来京留学。2017 年 9 月，学校成为北京市首批“一带一路”国家人才培养基地，获得了专项资金，进行国际建筑土木工程师人才培养和学科建设。学校还依托高精尖创新中心平台聘请了多位知名国际专家驻校讲学，派出“教师能力提升团组”赴外进修，带动学校科研合作水平和教师整体能力提升，这些工作为成立“一带一路”建筑类大学国际联盟奠定了良好的基础。

为进一步推动“一带一路”建筑类大学国际联盟可持续发展，张爱林提出四点建议：第一，建立“一带一路”建筑类大学国际联盟可持续发展与管理运行机制；第二，以信息化推动联盟及成员单位的资源开放共享；第三，设立开放基金，推进联盟成员协同创新研究；第四，建立“一带一路”建筑

图4 联盟首届轮值主席、北京建筑大学校长张爱林

类大学国际联盟校长论坛成果落地机制。张爱林说，让我们共同把“一带一路”倡议变成行动，把“一带一路”愿景变成现实，为建设更加和谐宜居的地球、建设更加美好的世界培养更多的创新人才，做出更大的贡献，共同铸就“一带一路”发展的新丰碑。

亚美尼亚国立建筑大学校长加吉克·加斯蒂安先生与沈阳建筑大学校长石铁矛先生分别代表中外院校代表致辞。他们向参加“一带一路”建筑类大学国际联盟成立大会的来宾表示热烈的欢迎，向为了此次大会的顺利召开做了大量筹备工作的北京建筑大学的各位同仁表示由衷的感谢。

加斯蒂安校长讲到，“一带一路”地区各国经济的发展，对沿线国家和地区建筑类高校提出了更高的要求和挑战，需要大家能够加强建造未来城市的能力。因此，大家应该聚焦三个主要目标，即加强人才的创新能力、在人才培养过程中融入安全可持续发展的理念，推动智慧城镇建设，共同推动资源的共享、环境的保护。北京建筑大学和亚美尼亚国立大学早在9年前就举行了第一届建筑与土木热点国际问题研讨会，针对城市建设领域当中的热点问题进行探讨，他特别感谢北京建筑大学所做出的努力。他提议，联盟成员建立夏季学校机制，交流各大学发展的技术和理念，并且分享建筑和工程方面的专业经验。最后，他建议各成员校利用现有的资源和积累的经验，共同推出区域产学研实验室，共同携手来为社会和经济问题提供解决方案。（图5）

图 5 亚美尼亚国立建筑大学校长加吉克·加斯蒂安

石铁矛校长认为，作为建筑类大学，各院校有着相近的发展现状和特点，在“一带一路”建设发展上，有着天然的学科和专业优势，这是大家开展合作交流、实现互利共赢的坚实基础。通过成立联盟，大家能够共同探索建筑类人才跨文化培养与跨境流动的新型培养模式，促进联盟高校间师生互动，更好地致力于科技创新、人才培养、文化交流，促进“一带一路”地区的共同发展。(图 6)

图 6 沈阳建筑大学校长石铁矛

葛巨众副主任致辞。他代表北京市教育委员会对大会的召开表示热烈祝贺。(图 7) 他谈到，北京正在积极打造中国的政治中心、文化中心、国际交往中心和科技创新中心，推动京津冀协同发展，向着国际一流的和谐宜居之都的目标迈进。为此，北京市教委将进一步加大教育开放的力度，支持北京高校与不同地区的高校交流互鉴，共同提升。“一带一路”建筑类大学国际联盟的成立正是适应了这一经济发展的新趋势和人类多样文明和谐共生的新潮流，能够为“一带一路”沿线国家和地区大学搭建教育信息共享和学术资源交流合作平台，探索跨国培养与跨境交流的人才培养新机制。他希望通过“一带一路”建筑类大学国际联盟这个平台，各国高校能够与北京的高校建立更加紧密的联系和更加深入的合作，共同应对全球化背景下高等教育面临的新挑战，共同促进各方的人才培养、科学研究和教育事业发展。

图 7　葛巨众副主任

最后，出席成立大会的 15 所中外院校代表集体签约、热情握手，共同祝贺“一带一路”建筑类大学国际联盟的成立。（图 8）

图 8　“一带一路”建筑类大学国际联盟在北京建筑大学成立

当天下午举行的校长论坛上，北京建筑大学张爱林校长以“把准未来导向，创新合作机制，培养建筑类创新人才”为题，做了精彩报告，美国夏威夷太平洋大学校长约翰·乔坦达、保加利亚建筑土木工程和大地测量大学校长伊万·马尔可夫、韩国大田大学校长李钟瑞、沈阳建筑大学校长石铁矛、吉林建筑大学校长戴昕、俄罗斯建筑土木科学院院士帕维尔·阿基莫夫、法国马恩·拉瓦雷大学副校长弗里德里克·杜马泽、法国英科工程大学联盟副校长文森特·希克斯、安徽建筑大学校长方潜生、马来西亚大学联盟主席拿督斯里·黄子炜、山东建筑大学副校长范存礼、河北建筑工程学院校长师涌江分别发言。与会代表紧密围绕“一带一路”倡议下建筑类大学国际交流合作与创新人才培养的主题进行深入探讨和交流，大家畅所欲言、相互借鉴，深入全面介绍各学校办学特色、国际交流与合作的成绩和经验，并对联盟建设和可持续发展提出了许多很好的建议。

图 9　韩国大田大学校长李钟瑞

图 10　美国夏威夷太平洋大学校长约翰·乔坦达

图 11　保加利亚建筑土木工程和大地测量大学校长伊万·马尔可夫

图 12　沈阳建筑大学校长石铁矛

图 13　俄罗斯建筑土木科学院院士
帕维尔・阿基莫夫

图 14　吉林建筑大学校长戴昕

图 15　法国英科工程大学联盟副校长
文森特・希克斯

图 16　法国马恩・拉瓦雷大学副校长
弗里德里克・杜马泽

图 17　安徽建筑大学校长方潜生

图 18　马来西亚大学联盟主席
拿督斯里・黄子炜

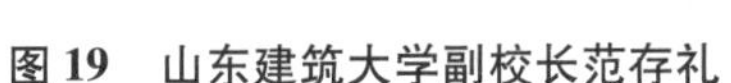
图 19　山东建筑大学副校长范存礼

图 20　河北建筑工程学院校长师涌江

大会闭幕式上，联盟秘书长和首任轮值主席、北京建筑大学校长张爱林发表了总结讲话。他说，一天紧张的大会开得热烈、友好、融洽、高效，收获了丰硕成果，达成了广泛共识。一是大家高度评价了“一带一路”倡议为建筑类大学发展、建筑类学科建设以及人才培养提供的重大机遇，并就携手共同紧紧把准机遇达成高度共识；二是大家一致认为“一带一路”建筑类大学国际联盟成立是建筑高等教育的内涵发展和特色发展，主动适应和服务区域经济发展的重要载体和平台，促进资源的开放与共享、互联互通，进一步推动联盟内建筑类大学在广度和深度的合作；三是大家纷纷表示联盟内建筑类大学将本着开放创新、务实合作的工作理念，以实际行动来推动和深化各领域合作不断取得新成果。他希望站在新起点上，大家共同携起手来，开拓进取，用实际的行动、“钉钉子”的精神，一步一个脚印地推动国际合作不断取得新成果。

下一届联盟会议将在马来西亚举行，联盟秘书长和首任轮值主席、北京建筑大学校长张爱林将会旗交给马来西亚大学联盟主席拿督斯里 · 黄子炜（图 21）。

大会闭幕后，北京建筑大学校长张爱林分别会见了美国夏威夷太平洋大学、保加利亚建筑土木工程和大地测量大学、亚美尼亚国立建筑大学、俄罗斯建筑土木科学院、韩国大田大学、法国马恩 · 拉瓦雷大学的相关代表，就联盟成立背景下校际下一步合作项目进行了深入交流。

图 21　会旗交接

图 22　“一带一路”建筑类大学国际联盟在北京建筑大学成立

“一带一路”建筑类大学国际联盟是按照“自愿平等、开放共享、合作共赢、创新发展”的原则自发组织的非营利性战略合作组织，旨在发挥特色优势、推进资源共享、加强协同创新、促进人才培养，提升国际科研合作水平，促进跨文化交流与理解，促进大学间人员、知识、技术等各方面资源的流动，培养高素质国际化人才，提升国际交流合作的能力与水平，服务经济结构转型升级，从而促进民心相通、跨国界产学研合作，提升大学内涵实力和国际声誉。联盟将致力于高素质、国际化工程技术人才培养，培育模式的创新实践，为社会培养高水平、实践能力强的工程技术人才，实现高层次工

程技术人才教育的新突破；致力于促进经济结构转型、产业升级，以集成解决社会共性关键技术为目标，以科研项目和技术创新为牵引，创新合作机制，打造跨国界多校对社会的协同创新平台，促进资金、产品、人才和服务的跨国界流动；致力于推进民心相通、跨文化交流与理解，积极促进大学间跨国界人员和文化交流，鼓励大学间人员跨国界流动，联合举办各类学术会议、科技竞赛、文化艺术及体育类活动，积极开展各项汉语推广活动。

图 23　在北京建筑大学图书馆门口合影

自 2017 年 5 月北京建筑大学发出倡议，截至目前，已有来自俄罗斯、波兰、法国、美国、英国、亚美尼亚、保加利亚、捷克、韩国、马来西亚、希腊、尼泊尔、以色列等 19 个国家的 44 所院校同意成立并加入联盟。其中既有安徽建筑大学、北京建筑大学、河北建筑工程学院、吉林建筑大学、山东建筑大学、沈阳建筑大学、天津城建大学、西安建筑科技大学等国内高校，也有亚美尼亚国立建筑大学、保加利亚建筑土木工程和大地测量大学、法国马恩·拉瓦雷大学、法国英科工程大学联盟集团、英国德比大学、希腊塞萨洛尼基亚里士多德大学、以色列贝扎雷艺术与设计学院、韩国大田大学、马来西亚理工大学、马来西亚大学联盟、尼泊尔工程学院、波兰华沙生态与管理大学、波兰华沙理工大学、莫斯科国立建筑学院、圣彼得堡国立建筑大学、俄罗斯建筑土木科学院、美国夏威夷太平洋大学等国外院校。

10 月 9 日晚，联盟举行预备会，与会代表审议了联盟章程、会徽（LOGO）及合作意向书，讨论了联盟可持续发展的运行机制及下一届联盟会议举办地点，一致同意北京建筑大学为“一带一路”建筑类大学国际联盟秘书长单位和首任轮值主席单位，北京建筑大学校长张爱林为联盟秘书长和首任轮值主席。（见图 24）

图 24　“一带一路”建筑类大学国际联盟成立大会预备会

Belt and Road Architectural University International Consortium Established at Beijing University of Civil Engineering and Architecture

On the morning of October 10th, the Inauguration Ceremony of the Belt and Road Architectural University International Consortium (BRAUIC) & Presidents'Forum were grandly held in the Jianben Lecture Hall at the Library of Daxing Campus, Beijing University of Civil Engineering and Architecture (BUCEA). Over 100 participants including Ge Juzhong, Deputy Director of Beijing Municipal Education Commission, presidents and representatives from 25 domestic and foreign architectural universities, as well media representatives had jointly witnessed the official establishment of the consortium, and conducted in-depth discussions and exchanges on the international exchange and cooperation of architectural universities and the cultivation of innovative talents under the Belt and Road Initiative.

Wang Jianzhong, Secretary of the CPC Committee of the BUCEA, presided over the Inauguration Ceremony and delivered a welcome speech. He said that, education plays a fundamental and leading role in the joint construction of the Belt and Road and is an important part of it. The proposition of the great Belt and Road Initiative has presented us with major opportunities for improving the educational levels of countries and regions along the Belt and Road, facilitating people-to-people and cultural exchanges and promoting regional prosperity and development. Facing the major historical opportunities for infrastructure interconnection in the construction of the Belt and Road and the urgent need for architectural talents and technologies, it is the bounded historical mission for architectural universities to found the BRAUIC, so as to promote the exchange and cooperation of architectural universities along the Belt and Road, innovate the cultivation mechanism of architectural talents, explore the development of cross-cultural architectural disciplines, push forward the commercialization of scientific and technological achievements, serve the development and construction of countries and regions along the Belt and Road and in the Eurasian region, and be fully devoted to and actively integrated into this great practice, which is of great significance to promoting common prosperity of the Belt and Road education. The founding of the BRAUIC marks the very first solid step in the cooperation of architectural universities along the route; we will further intensify pragmatic cooperation, strive for mutual benefit and win-win results and establish more cooperation and exchange mechanisms, thus making even greater contributions to the prosperity and development of urban and architectural undertakings around the world, as well as to the harmonious and livable well-being of mankind.

Under the theme of "Bring the Initiative into Actions, Transform the Vision into Reality, Innovatively Promote International Exchanges and Cooperation Among the Architectural Universities along the Belt and Road", Zhang Ailin, the 1st Rotating Chairman of the BRAUIC and President of BUCEA made a report on the background, charter and cooperation intention of the consortium. He said that in the past couple of years, BUCEA has carried out extensive cooperation with countries along the Belt and Road, developed partnerships with a group of universities, signed inter-university cooperation agreements, arranged exchanges of teachers and students with them, carried out scientific research cooperation, co-hosted international conferences and opened a new chapter in exchanges and cooperation. In 2016, BUCEA won a special scholarship from the Beijing Municipal Education Commission arranged for funding aspiring young students from countries along the Belt and Road to study in Beijing. In this September, BUCEA became one of the first batch of Belt and Road national talent cultivation bases in Beijing, and received special funds for the cultivation of international architectural and civil engineering talents and discipline construction. Based on the platform of the high-tech innovation center, BUCEA has also invited many well-known international experts to give lectures in China, and dispatched the "team for enhanced teaching capability" to study abroad, which has elevated its level of scientific research cooperation, and improved the overall ability of its teaching faculty. Such efforts have laid down a sound foundation for the founding of the BRAUIC.

In order to further promote the sustainable development of the consortium, Zhang Ailin put forward four suggestions: firstly, establish a sustainable development and management operation mechanism of the BRAUIC; secondly, promote open and shared resources among the BRAUIC members through information technologies; thirdly, set up an open fund for collaborative innovation research among the members; fourthly, establish a mechanism for implementing the achievements concluded in the BRAUIC Presidents'Forum. He called for everyone to work together to turn the Belt and Road Initiative into action and the Belt and Road vision into reality, cultivate more innovative talents and make more contributions to building a more harmonious, livable and better world, and jointly create a new monument for the Belt and Road development.

Mr. Gagik Galstyan, President of National University of Architecture and Construction of Armenia, and Mr. Shi Tiemao, President of Shenyang Jianzhu University delivered speeches on behalf of Chinese and foreign universities respectively, in which they extended a warm welcome to the guests who participated in the BRAUIC inauguration ceremony, and expressed their heartfelt gratitude to the BUCEA staff for their preparatory work for the smooth convention of the conference and forum.

President Galstyan said that, the economic development of all countries along the Belt and Road has imposed higher requirements and challenges on architectural universities of these countries, and the ability to build future cities needs to be further enhanced; therefore we should focus on three main objectives, i. e. strengthening the innovative ability of talents, integrating the concept of safe and sustainable development into the process of talent training, promoting the construction of smart cities and towns, and jointly facilitating resource sharing and environmental protection. As early as nine years ago, his university and BUCEA had co-held the first international seminar on hot topics in architecture and civil engineering, which mainly discussed on hot issues in the field of urban construction, and for this he expressed special gratitude for the efforts made by BUCEA. He proposed that the BRAUIC members should launch a summer school mechanism to exchange technologies and ideas developed by each university, and share professional experiences in architecture and engineering. He proposed in the end that the BRAUIC members should jointly set up regional industry-university-research laboratories by availing existing resources and accumulated experiences, and work together to provide solutions to social and economic issues.

President Shi Tiemao said that, architectural universities have similar development status and features, and possess academic and professional advantages for the Belt and Road construction, which serves as a solid foundation for cooperation and communication with mutual benefit and win-win results. Through the consortium we could jointly explore the new cultivation model of cross-cultural and cross-border architectural talents, step up the interaction of teachers and students among the member universities, better commit to technological innovation, talent training, cultural exchange, and promote common development in the Belt and Road regions.

On behalf of the Beijing Municipal Education Commission, Ge Juzhong, Deputy Director delivered a speech and expressed congratulations to the convening of the conference. He said that, Beijing is now actively building itself into China's political center, cultural center, international exchange center and technological innovation center, promoting the coordinated development of Beijing, Tianjin and Hebei, and marching towards the goal of becoming a world-class harmonious and livable city. To this end, the Beijing Municipal Education Commission will further open up its education, support mutual exchanges between universities in Beijing and other regions and seek for common progress.

The establishment of the BRAUIC corresponds to the new trend of economic development and harmonious coexistence of diversified civilizations, and can serve as a platform for sharing education resources and academic exchanges and cooperation, and exploring new mechanisms for cross-border talent cultivation and exchange. He hopes that through the BRAUIC platform, universities in various countries could be bonded even closer and conduct deeper cooperation with universities in Beijing, so as to jointly address the new challenges for higher education in the context of globalization, push forward talent cultivation, scientific research and educational development for all sides.

In the end, representatives of 15 Chinese and foreign universities attending the inaugural ceremony signed a collective agreement and shook hands to congratulate the founding of the BRAUIC.

At the Presidents'Forum held in the afternoon, President Zhang Ailin of BUCEA delivered a report themed on Hold the Future Orientation, Innovate the Cooperation Mechanism and Cultivate Innovative Talents in Architecture. Speeches were also given by John Gotanda, President of Hawai'i Pacific University; Ivan Markov, President of University of Architecture, Civil Engineering and Geodesy of Bulgaria; Lee Jong Seo, President of Daejeon University of South Korea; Shi Tiemao, President of Shenyang Jianzhu University; Dai Xin, President of Jilin Jianzhu University; Pavel Akimov, Academician of Academy of Architecture and Civil Engineering of Russia, Frederick Toumazet, Vice President of Université de Marne-La-Vallée of France; Vincent Six, Vice President of Yncrea of France; Fang Qiansheng, President of Anhui Jianzhu University; Dato'Seri Wong Tze Wei, Chairman of Consortium of Malaysian University; Fan Cunli, Vice President of Shandong Jianzhu University, and Shi Yongjiang, President of Hebei University of Architecture. Delegates had made in-depth and free discussions and exchanges on international exchange, cooperation and innovative talent training in architectural universities under the Belt and Road Initiative, introduced in detail the situations of their respective universities, shared their achievements and experiences in international exchanges and cooperation, and put forward multiple good suggestions for the BRAUIC construction and sustainable development.

In the closing ceremony, Zhang Ailin, Secretary-General of and the 1st Rotating Chairman of the BRAUIC, President of BUCEA made a concluding speech. He said that, fruitful results and extensive consensus had been reached after a day's conference of passionate, friendly, harmonious and efficient exchanges; firstly, everyone highly complimented the Belt and Road Initiative for the major opportunities coming along with it for the development of architectural universities, architectural discipline construction and talent cultivation, and reached unanimous consensuses on working together to size tight such opportunities; secondly, everyone agreed that the BRAUIC is an important carrier and platform for architectural higher education to develop its contents and characteristics, and actively adapt to and serve regional economic development; the consortium would promote the opening and sharing of resources and interconnection, and further step up cooperation of the BRAUIC member universities in a wider and deeper scope; thirdly, everyone expressed that the architectural universities in the consortium will follow the working concept of open, innovation and pragmatic cooperation, and make earnest efforts in promoting and deepening cooperation in various fields for new results. He hopes that, standing at a new starting point, we shall join hands to forge ahead, exert unswerving efforts to earnestly promote international cooperation for achieving new results.

As the next session of the BRAUIC conference will be held in Malaysia, Zhang Ailin, Secretary-general & the 1st Rotating Chairman of the BRAUIC and President of BUCEA, handed over the flag to Dato'Seri Wong Tze Wei, Chairman of Consortium of Malaysian University.

After the closing ceremony, Zhang Ailin, President of BUCEA met with representatives from Hawai'i Pacific University, University of Architecture, Civil Engineering and Geodesy of Bulgaria, National University of Architecture and Construction of Armenia, Academy of Architecture and Civil Engineering of Russia, Daejeon University of South Korea and Université de Marne-La-Vallée of France, and conducted in-depth exchanges on the next step of inter-school cooperation projects after the establishment of the BRAUIC.

As a non-profit strategic cooperation organization founded on the principles of "voluntary & equal, open & sharing, win-win cooperation and innovative development", the BRAUIC aims to play its unique advantages, promote resource sharing, enhance coordinated innovation, facilitate talent cultivation, elevate the level of international scientific research cooperation, facilitate cross-cultural exchanges and understanding, step up the flow of resources in all aspects of personnel, knowledge, technology among universities, cultivate high-quality international talents, improve the capability of international exchanges and cooperation, serve economic structural transformation and upgrading, thus strengthening people-to-people ties and cross-border industry-university-research cooperation, and improving the strength and international reputation of the universities. The BRAUIC will be committed to cultivating high-quality, international engineering and technical personnel, carrying out innovative educational practices, providing the society with high-level engineering and technical personnel with strong practice skills, and achieving new breakthroughs in high-level engineering and technical personnel education. The BRAUIC will be engaged to push forward economic structural transformation and industrial upgrading, aim at providing integrated solutions of key technologies for the common demand of the society, be driven by scientific research projects and technological innovation, innovate cooperation mechanisms, create a coordinated innovation platform between multiple cross-border universities and societies, facilitate the flow of funds, products, talents and services across borders. The BRAUIC will also work to boost people-to-people communication, cross-cultural exchanges and understanding, actively promote cross-border personnel and cultural exchanges among universities, encourage cross-border mobility of inter-university personnel, jointly organize academic conferences, scientific and technological competitions, cultural, art and sport activities, and actively launch Chinese language activities.

Since this May when BUCEA launched this initiative, 44 universities from 19 countries have agreed and joined the consortium, including Russia, Poland, France, the United States, the United Kingdom, Armenia, Bulgaria, the Czech Republic, South Korea, Malaysia, Greece, Nepal and Israel, with domestic universities such as Anhui Jianzhu University, BUCEA, Hebei University of Architecture, Jilin Jianzhu University, Shandong Jianzhu University, Shenyang Jianzhu University, Tianjin Chengjian University, Xi'an University of Architecture and Technology, as well as foreign universities such as National University of Architecture and Construction of Armenia, University of Architecture, Civil Engineering and Geodesy of Bulgaria, Université de Marne-La-Vallée of France, Yncrea of France, University of Derby of the United Kingdom, Aristotle University of Thessaloniki, Greece, Bezalel Academy of Art and Design of Israel, Daejeon University of South Korea, University of Technology Malaysia, Consortium of Malaysian University, Nepal Engineering College, University of Ecology and Management in Warsaw of Poland, Warsaw University of Technology of Poland, Moscow State University of Civil Engineering, St. Petersburg State University of Architecture, Academy of Architecture and Civil Engineering of Russia, and Hawai'i Pacific University of the United States of America.

On the evening of October 9, the BRAUIC held a preparatory meeting in which the participants reviewed the consortium's charter, logo and letter of intent, discussed the operation mechanism for the consortium's sustainable development and the location for the next consortium conference, and unanimously chosen BUCEA as the Secretary-General unit and the 1st rotating chairman unit, with Zhang Ailin, President of BUCEA as Secretary-General and the 1st Rotating Chairman.

北京建筑大学校长张爱林率团赴马来西亚主持“一带一路”建筑类大学国际联盟2018年会议暨校长论坛

2018年9月25日至27日，作为“一带一路”建筑类大学国际联盟主席及2018年会议暨校长论坛组织委员会主席，北京建筑大学校长张爱林率团赴马来西亚吉隆坡，主持“一带一路”建筑类大学国际联盟2018年会议暨校长论坛，先后与亚美尼亚国立建筑大学、保加利亚土木建筑及大地测量大学、马来西亚博特拉大学、马来西亚国立大学、马来西亚理工大学等高校商谈合作事宜。

图1　“一带一路”建筑类大学国际联盟2018年会议暨校长论坛

9月27日，2018年会议暨校长论坛在马来西亚理工大学隆重召开。（图1）来自国内外13所建筑类大学校长和代表共计50余人与会，共同研讨推进“一带一路”建筑类大学国际联盟建设，进一步创新务实合作，并围绕“一带一路”建筑类创新人才培养及“一带一路”沿线国家城市基础设施建设与建筑遗产保护两个主题展开深入交流。

2018 年会议暨校长论坛组织委员会执行主席、马来西亚大学联盟主席黄子炜主持联盟 2018 年会议并致欢迎辞。（图 2）他首先对北京建筑大学在主办此次会议中所做的各项工作和辛苦付出表示感谢。他讲到，此次联盟会议暨校长论坛落地吉隆坡，他感到十分荣幸，希望这次会议将推动“一带一路”沿线建筑类大学交流合作，推进不同国家间建筑技术与信息交换，服务“一带一路”沿线地区的发展建设。

图 2　2018 年会议暨校长论坛组织委员会执行主席、马来西亚大学联盟主席黄子炜

联盟主席、2018 年会议暨校长论坛组织委员会主席、北京建筑大学校长张爱林（图 3）以“用务实的行动建设‘一带一路’建筑类大学国际联盟”为题发表讲话，首先总结了联盟建立近一年来的工作，就如何创新推动联盟可持续发展提出了三点建议。第一，进一步完善联盟日常工作机制。建议设立副主席单位及常务理事单位，成员单位专人负责日常工作推进，同时建设“一带一路”建筑类大学国际联盟网站，实现信息互联互通、资源开放共享。第二，以培养通晓和掌握“一带一路”沿线国家国际规则的建筑类高质量创新型人才为抓手，发挥各自国家和区域的政策优势以及联盟成员单位的学科专业优势，研讨构建联盟内开放式、国际化教育教学体系。第三，构建联盟成员高校教师交流访问合作的工作机制，引导共同开展科学研究，提升师资队伍国际化水平。

图 3　联盟主席、2018 年会议暨校长论坛组织委员会主席、北京建筑大学校长张爱林

亚美尼亚国立建筑大学校长加吉克·加斯蒂安代表联盟高校致辞（图 4）。他充分肯定了联盟对“一带一路”沿线建筑类大学交流合作的重要意义及成立以来良好的工作基础。他指出，要想将“一带一路”倡议落到实处，必须从公路、桥梁、铁路、隧道、建筑工程、旅游景点六大“通道工程”着手。因此，他建议联盟成员立足现有合作基础，发挥最大优势，围绕教育交流、科研合作及行政管理三个方面开展务实合作，一步一个脚印推进联盟合作产出更为丰硕的成果。

图 4　亚美尼亚国立建筑大学校长加吉克·加斯蒂安

马六甲伊斯兰大学学院副校长巴哈鲁迪安代表马来西亚高校致辞（图5）。他谈到，“一带一路”建筑类大学国际联盟是提高沿线国家和地区教育水平、推动各高校人文交流与教育发展的重要平台，为马来西亚大学国际化发展与全球化人才培养提供了重大机遇。

图 5　马六甲伊斯兰大学学院副校长巴哈鲁迪安

当天下午举行的校长论坛上，北京建筑大学校长张爱林以“把准‘两大服务’定位，建设创新型建筑类学科”为题，做了精彩报告（图 6），保加利亚建筑土木工程和大地测量大学校长伊万·马尔可夫（图 7）、山东建筑大学国际交流合作处处长陈宝明（图 8）、马来西亚博特拉大学设计与建筑学院副院长约哈里（图 9）、马来西亚测绘大学学院副院长纳琪拉（图 10）、马来西亚国立大学副教授阿兹兰（图 11）、英迪国际大学高级讲师艾利克斯（图 12）分别发言，与会代表紧密围绕“一带一路”建筑类创新人才培养及“一带一路”沿线国家城市基础设施建设与建筑遗产保护两个主题进行深入的探讨和交流。大家畅所欲言、相互借鉴，深入全面介绍各学校办学特色、国际交流与合作的成绩和经验，积极为联盟建设和可持续发展建言献策。

图 6　北京建筑大学校长张爱林

图 7　保加利亚建筑土木工程和大地测量大学校长伊万・马尔可夫

图 8　山东建筑大学国际交流合作处处长陈宝明

图 9　马来西亚博特拉大学设计与建筑学院副院长约哈里

图 10　马来西亚测绘大学学院副院长纳琪拉

图 11　马来西亚国立大学副教授阿兹兰

图 12　英迪国际大学高级讲师艾利克斯

会议闭幕式上，联盟主席、北京建筑大学校长张爱林总结讲话。他谈到，"一带一路"建筑类大学国际联盟是在积极响应、落实中国"一带一路"倡议的时代背景下应运而生，是顺应世界多极化、经济全球化及教育国际化的产物。未来我们将坚持共建共享、合作共赢、交流互鉴，开展广泛性的常态合作交流，携起手来，一步一个脚印地推进各项任务目标的实施，一点一滴抓出建设成果，推动联盟走深走实，实现高质量建设发展。

会议期间，联盟主席、北京建筑大学校长张爱林与参会联盟高校校长展开深入交流，共商发展之策，围绕如何健全联盟组织机构与体制建设、创新联盟高校合作方式与机制、培养国际化人才、服务"一带一路"沿线国家和地区城乡建设与发展等进行深入研讨，对联盟未来发展方略及行动计划达成共识。

2018 年 9 月 26 日下午，联盟举行 2018 年会议暨校长论坛预备会，联盟主席、北京建筑大学校长张爱林向各位大学代表作联盟工作报告，提出下一年联盟工作计划。与会代表审议了申请加入联盟院校的基本情况，一致同意俄罗斯国立莫斯科建筑大学、挪威卑尔根建筑学院、印度科技教育集团、吉尔吉斯斯坦奥什科技大学、美国里海大学、塞尔维亚诺维萨德大学 6 名新成员加入联盟（图 13）。

图 13 “一带一路”建筑类大学国际联盟 2018 年会议暨校长论坛预备会

President Zhang Ailin of Beijing University of Civil Engineering and Architecture Chairs the 2018 Belt and Road Architectural University International Consortium Conference & Presidents' Forum in Malaysia

From September 25th to 27th, as Chairman of the Belt and Road Architectural University International Consortium (BRAUIC) and Chairman of the 2018 BRAUIC Conference & Presidents' Forum Organizing Committee, Zhang Ailin led a delegation to Kuala Lumpur, Malaysia to preside over the 2018 BRAUIC Conference & Presidents' Forum, and discuss cooperation with National University of Architecture and Construction of Armenia, University of Architecture, Civil Engineering and Geodesy of Bulgaria, Universiti Putra Malaysia, Universiti Kebangsaan Malaysia, National University of Malaysia, University of Technology Malaysia, etc.

On September 27th, the 2018 BRAUIC Conference & Presidents' Forum was grandly held at University of Technology Malaysia. Over 50 presidents and representatives from 13 domestic and foreign construction universities attended the event and discussed on promoting the BRAUIC construction, intensifying innovative and pragmatic cooperation, and made in-depth exchanges under the themes of cultivating Belt and Road innovative architectural talents as well as urban infrastructure construction and architectural heritage protection in countries along the Belt and Road.

Dato' Seri Wong Tze Wei,, Executive Chairman of the 2018 BRAUIC Conference and Presidents' Forum Organizing Committee, Chairman of Consortium of Malaysian University, presided over the conference and delivered a welcome speech. He expressed his gratitude to Beijing University of Civil Engineering and Architecture (BUCEA) for all the hard preparatory work for the event, felt very honored for this conference and forum being held in Kuala Lumpur, and hoped that this event would promote the exchange and cooperation of architectural universities along the Belt and Road, step up exchanges of architectural technologies and information among different countries, and serve the development of Belt and Road countries and regions.

Zhang Ailin, Chairman of the 2018 BRAUIC Conference & Presidents' Forum Organizing Committee and President of BUCEA, delivered a speech on the topic of "Build the Belt and Road Architectural University International Consortium with Concrete Actions". In his speech, he put forward three suggestions on innovating and promoting the sustainable development of the consortium: firstly, further improving the consortium's daily working mechanism; he proposed to establish a vice-chairman unit and an executive director unit, with dedicated members in charge of routine work, and establish the BRAUIC website for information interconnection and resource sharing; secondly, cultivating high-quality and innovative architectural talents who know well about the international rules of the Belt and Road countries, availing policy advantages of respective countries and regions, as well as academic and professional advantages of the BRAUIC members, and research on establishing an open and international educational system within the consortium; thirdly, setting up a working mechanism for the exchange and cooperation of teachers from the consortium, guiding the joint development of scientific research and enabling the teaching faculty to be more internationalized.

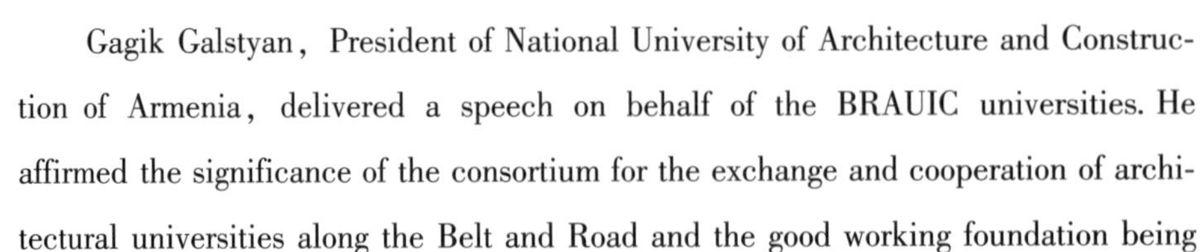

Gagik Galstyan, President of National University of Architecture and Construction of Armenia, delivered a speech on behalf of the BRAUIC universities. He affirmed the significance of the consortium for the exchange and cooperation of architectural universities along the Belt and Road and the good working foundation being laid since its establishment, and pointed out that in order to practically implement the Belt and Road initiative, we shall start with the six "passage" projects of highways, bridges, railways, tunnels, construction and tourist attractions. He also proposed that members of the consortium should, basing on the existing cooperation, give full play to their respective advantages, conduct practical cooperation in educational exchange, scientific research and administrative management, and earnestly strive for more abundant results.

Baharudian, Vice President of Malacca Islamic University, delivered a speech on behalf of Malaysian universities. He said that the BRAUIC is an important platform for improving the educational level and promoting personnel exchanges and educational development of universities in the Belt and Road countries and regions; it also presents major opportunities for the international development of Malaysian universities and cultivation of globalized talents.

At the presidents' forum held in the afternoon, Zhang Ailin, President of BUCEA made a report on the topic of "Positioning the Two Major Services and Establishing Innovative Architectural Discipline" . Speeches were also given by Ivan Markov, President of University of Architecture, Civil Engineering and Geodesy of Bulgaria; Chen Baoming, Director of the International Exchange and Cooperation Office of Shandong Jianzhu University; Johari, Deputy Dean of the College of Design and Architecture, Botra University, Malaysia; Nakira, Vice President of Malaysia University of Surveying and Mapping; Azlan Alix, Associate Professor of Universiti Kebangsaan Malaysia, senior lecturer of INTI International University. Delegates made deep and free discussions and exchanges on the topics of innovative talent cultivation for the Belt and Road construction as well as urban infrastructure construction & architectural heritage protection along the Belt and Road; they also introduced in detail the characteristics of their universities, shared the achievements and experiences in international exchanges and cooperation, and actively contributed ideas for the construction and sustainable development of the consortium.

At the closing ceremony, Zhang Ailin, Chairman of the BRAUIC and President of BUCEA said in his speech that the BRAUIC came into being in response to the implementation of China's Belt and Road Initiative and to global multi-polarization, economic globalization and education internationalization. In the future, we will adhere to joint construction and sharing, win-win cooperation, exchanges and mutual observation, carry out extensive cooperative exchanges, and work together to realize various goals step by step, yield results bit by bit, and push forward intensive, pragmatic and high-quality development of the consortium.

During the meeting, Zhang Ailin, Chairman of the BRAUIC and President of BUCEA conducted in-depth exchanges with the presidents of the participating universities to discuss on development strategies, improving the organization and system construction of the consortium, innovating the cooperation methods and mechanisms of universities, cultivating international talents, and serving urban and rural construction and development in countries and regions along the Belt and Road. Upon that, consensuses have been reached on the future development strategies and action plans for the BRAUIC.

On the afternoon of September 26th, the preparatory meeting for the 2018 BRAUIC Conference and Presidents' Forum was held, in which Zhang Ailin, Chairman of the BRAUIC and President of BUCEA reported to university representatives on the work of the consortium and proposed the consortium work plan for the next year. Delegates reviewed and deliberated the basic situation of universities applying for joining the consortium, and unanimously agreed the recruitment of six new members, i. e. the Moscow State University of Civil Engineering of Russia, Bergen School of Architecture of Norway, Techno India Group, Osh University of Technology of Kyrgyzstan, Lehigh University of the United States, and Novi Sad University of Serbia.

四

校长论坛报告

Presidents' Report

在“一带一路”建筑类大学国际联盟成立大会上的致辞

北京市教育委员会副主任　葛巨众

尊敬的各位校长、各位来宾、女士们、先生们，大家上午好！非常荣幸与各位嘉宾欢聚一堂，共同庆祝“一带一路”建筑类大学国际联盟成立。我代表北京市教育委员会，对大会的召开表示热烈的祝贺！对各位来宾的到来表示诚挚的欢迎！

北京是一座具有首都风范、古都风韵、时代风貌的美丽城市。历史上的北京是中国的六朝古都，有三千多年的建城史、八百多年的建都史，历史悠久、文化深厚。这里有长城、故宫、天坛等文明古迹，有京剧、大鼓、评书等传统艺术，它们是北京的历史印记和文化名片，谱写着这座古城独特的风韵。这里居住着中国的56个民族和来自全国各地的人们，他们的生活习惯、语言和风俗在长期交往中互相融合、互相渗透，形成了北京包容、开放、多元的人文特色。今天的北京正在积极打造中国的政治中心、文化中心、国际交往中心和科技创新中心，推动京津冀协同发展，向着国际一流和谐宜居之都的目标迈进。为此我们将进一步加大教育开放的力度，支持北京高校与不同地区的高校交流互鉴，共同提升。随着经济全球化的深入发展，和平、发展、合作、共赢的时代潮流更加强劲，各国日渐成为“你中有我，我中有你”的命运共同体。在追求自身利益时，要兼顾他方利益；在寻求自身发展时，要促进共同发展，必须坚持不同文明兼容并蓄、交流互鉴。今天“一带一路”建筑类大学国际联盟的成立正是顺应了这一经济发展的新趋势和人类多样文明和谐共生的新潮流，能够为“一带一路”沿线国家和地区大学搭建教育信息共享、学术资源交流的合作平台，探索跨国培养与跨境交流的人才培养新机制。

我衷心希望，各联盟成员在今后的交流中，能够推进协同创新，加强联盟高校间教学科研合作及师生的交流。坚持相互尊重、平等相待的原则，求同存异、共同发展，推动不同文明的交流互鉴与和谐共生。

北京是中国的首都，北京市民正在和全国人民一道，为实现“两个一百年”的奋斗目标和实现中华民族伟大复兴的中国梦而努力奋斗。中国发展目标的实现，需要与世界各国加强合作、取长补短；北京高等教育的发展，也需要与各方互学互鉴、共同提升。我希望通过“一带一路”建筑类大学国际联盟这个平台，各国各地区高校能够与北京的高校建立更加紧密的联系和更加深入的合作，共同应对全球化背景下高等教育面临的新挑战，共同促进各方的人才培养、科学研究和教育事业发展。

最后，预祝“一带一路”建筑类大学国际联盟成立大会圆满成功！祝各位来宾在北京访问愉快。谢谢大家！

Speech on the Belt and Road Architectural University International Consortium Inauguration Conference

Ge Juzhong, Deputy Director of Beijing
Municipal Education Commission

Honorable presidents, guests, ladies and gentlemen,

Good morning! I am much honored to get together with all guests here today to jointly congratulate the establishment of the Belt and Road Architectural University International Consortium. First of all, I show my warmly welcome to the opening of the conference and to the coming of all guests on behalf of Beijing Municipal Education Commission.

Beijing is a beautiful city with capital demeanour, the charm of ancient capital and contemporary spirits. In history, Beijing is the ancient capital of six dynasties of China, and it has over 3,000 years of activating history and over 800 years of capital history. It has a long profound cultural history. There are many historical sites, such as the Great Wall, Imperial Palace and Temple of Heaven, etc. and there are many traditional arts, such as Peking Opera, Bass Drum and Pingshu, etc. They are historical imprint and cultural name cards of Beijing, composing the unique charm of this ancient city. Fifty-Six nations of people from all regions of China are living there, and their living habits, languages and customs integrate and interpenetrate mutually in long-term communication and has formed the forgive, the open and diversified human feature of Beijing. At present, Beijing is actively becoming the political center, cultural center, international communication center and scientific innovative center of China, promoting the coordinated development of Beijing-Tianjin-Hebei Region. It is stepping towards the international first-class and livable capital. Therefore, we will further strengthen the education openness extent, support the communication and mutual identification among colleges and universities in Beijing and different regions to improve mutually. With the deep development, peach, development and cooperation of economic globalization, the win-win trend of the times will be stronger, and all countries will be open to each other and become a community of common future. While all countries are pursuing their own interests, they shall consider the other countries'interests at the same time. While they are searching for their own development, they shall promote joint development. They shall persist on integrating, communicating and mutually identifying different civilizations. Today, the Belt and Road Architecture University International Consortium is just adapted to the new trend of such economic development and new trend of harmonious coexistence of diversified human civilizations, which could establish a platform for sharing education information, communicating and cooperating the academic resources, exploring new system for multi-national cultivation and international communication talent cultivation for the national and regional universities along the Belt and Road.

The members of Consortium could jointly promote collaborative innovation, and strengthen the communication and cooperation among universities of consortium, teaching and scientific researches and teachers and students. I sincerely hope that in the future communication among all parties, we will persist on the principle of mutual respect and equal treatment, seek common points while reserving differences, develop jointly and promote the communication and mutual identification of different civilizations and coexistent harmoniously.

Beijing is the capital of China, and Beijing citizens are working hard together with all Chinese people to realize the two one-hundred years of goals and realize the Chinese dream of bringing about a great rejuvenation of the Chinese nation. The realization of China development target needs to strengthen cooperation with all other countries around the world and learn from the others' strong points to offset its own weakness. The development of higher education in Beijing also needs to learn from each other, mutually identify to improve jointly. I hope all colleges and universities of all countries could establish closer contract and deeper cooperation with the colleges and universities in Beijing through the platform of the Belt and Road Architectural University International Consortium, to jointly face to the new challenges of higher education under the background of globalization and jointly promote the development of talent cultivation, scientific research and education business of all parties.

Finally, I hope the Belt and Road Architectural University International Consortium Inauguration Conference will be successfully completely and hope every guest to enjoy a happy journey to Beijing. Thanks!

把倡议变为行动，把愿景变为现实，创新推进“一带一路”建筑类大学国际交流合作

北京建筑大学校长　张爱林

尊敬的北京市教委葛巨众副主任，尊敬的各位大学校长、各位嘉宾，女士们，先生们：

今天我们相聚在北京建筑大学，共同见证“一带一路”建筑类大学国际联盟的成立，共同研讨创新推进“一带一路”建筑类大学国际交流合作及创新人才培养。我谨代表北京建筑大学和联盟所有的发起单位，对大家积极支持成立“一带一路”建筑类大学国际联盟表示衷心的感谢！对大家出席今天的成立大会和校长交流会表示热烈的欢迎！

2013 年 9 月和 10 月，中国国家主席习近平在出访中亚和东南亚国家期间，提出了“一带一路”倡议，就是共建“丝绸之路经济带”和“21 世纪海上丝绸之路”倡议。2017 年 5 月，在北京成功举办了举世瞩目的“一带一路”国际合作高峰论坛，习近平主席提议将“一带一路”建成和平之路、繁荣之路、开放之路、创新之路和文明之路。与会各方对外发出了合力推动“一带一路”国际合作、携手共建人类命运共同体的积极信号，这一次高峰论坛进一步明确了未来“一带一路”国家的合作方向，规划了“一带一路”建设的具体路线图，确定了一批“一带一路”将实施的重点项目。

“一带一路”倡议有三个内涵要义：一是突出互联互通的 21 世纪特色，二是建设经济发展带，三是得道多助、互利多赢。“一带一路”是合作发展的理念和倡议，是和平合作、开放包容、互学互鉴、互利共赢的理念，为沿

线国家和地区带来共同的发展机会，拓展更加广阔的发展空间。

我们常说，“少年强则国家强”，各国青年的互鉴交流、开拓创新是实施“一带一路”倡议、推动“一带一路”可持续发展的重要力量。我们要坚持为互联网时代、未来一代的各国青年成就青春梦想，搭建更多合作平台，开辟更多合作渠道，推动高等教育实质国际合作，提升高等教育合作水平。

北京建筑大学作为北京地区唯一的建筑大学，是国家住建部和北京市的共建大学。北京建筑大学始建于1907年，2017年是建校110周年，建筑类学科具有悠久的历史。目前已与28个国家和地区63所大学建立长期合作关系，1999年与法国政府共建中法能源培训中心，拥有“北京未来城市设计高精尖创新中心”“代表性古建筑与古建筑数据库”教育部工程研究中心等26个省部级高端科研平台，特别是针对丝绸之路沿线建筑遗产保护开展了大量极具价值的研究和保护工作，先后完成了世界最大的线性遗产廊道项目——丝绸之路新疆地区古代寺庙遗址保护规划、长城保护规划、云冈石窟保护研究以及柬埔寨周萨神殿保护维修工程等多个世界文化遗产项目的保护规划等工作。

最近两年，我校与“一带一路”沿线国家高校开展了广泛合作，发展了一批新的友好伙伴高校，签署校际合作协议，双方进行师生交流、开展科研合作、共同举办国际会议，开启了交流合作的新篇章。我们主办了北京城市设计国际高峰论坛、建筑与土木热点问题国际研讨会、亚太城市建设与管理实务论坛、“空间数据基础设施建设与应用”国际研讨会、“一带一路”历史建筑摄影·手绘艺术展等。

2016年，我校获得北京市教委“一带一路”沿线国家留学生专项奖学金，用于资助“一带一路”沿线国家有志青年来京留学。2017年9月，我校成为北京市首批“一带一路”国家人才培养基地，获得专项资金进行国际建筑土木工程师人才培养和学科建设。我们还聘请了多位知名国际专家驻校讲学，派出“教师能力提升团组”赴外进修，带动学校科研合作水平和教师整体能力提升。这些工作为成立“一带一路”建筑类大学国际联盟奠定了良好的基础。

北京建筑大学面临难得的四个重大发展机遇，一是“一带一路”发展倡议加快“一带一路”基础设施建设；二是中国的建筑业向信息化、绿色化、工业化转型升级；三是首都北京新定位，建设世界一流的和谐宜居之都；四是京津冀协同发展，加快建设雄安新区。这要求我校努力建设国际知名的开放式、创新型建筑大学，要求我们推进与各个国家的大学全面交流与合作，推动教育综合改革和教育国际化进程，搭建国际化人才培养、科研协同创新及人文交流的开放共享平台。因此我们产生了成立“一带一路”建筑类大学国际联盟的意向，这立即得到了北京市政府教委、北京市外办的肯定和大力支持。我们发出了倡议书，起草了联盟章程，迅速得到了国内外各兄弟院校的积极响应和大力支持。从 2017 年 5 月份开始，我们陆续收到各兄弟院校加入联盟的确认函，截至目前，已有来自俄罗斯、波兰、法国、美国、英国、亚美尼亚、保加利亚、捷克、韩国、马来西亚、希腊、尼泊尔、以色列等 19 个国家的 44 所大学同意成立并加入联盟。今天，有 11 个国家 25 所大学的 51 位代表参会，共同见证“一带一路”建筑类大学国际联盟的正式成立。我们相信，随着联盟的进一步发展扩大，将有更多的中国高校和“一带一路”沿线其他国家的高校加入进来。

为推动“一带一路”建筑类大学国际联盟可持续发展，我们提出如下四点建议。

第一，建立“一带一路”建筑类大学国际联盟可持续发展与管理运行机制，建立开放基金制度。

第二，以信息化推动联盟及成员单位的资源开放共享。

第三，设立开放基金，推进联盟成员协同创新研究。在人才培养、科学研究、师资队伍建设、校园文化、社会服务等方面开展实质合作研究，促进不同学校优势特色学科的交叉融合。

第四，建立“一带一路”建筑类大学国际联盟校长论坛成果落地机制。

中国人有两句谚语，一句是“良好的开端等于成功的一半”，今天，“一带一路”建筑类大学国际联盟成立了，联盟的建设已经迈出坚实的第一步，有了良好的开端，我们充满信心；另一句是“行百里者半九十”，未来的路还很长，任务还很艰巨，我们要鼓足干劲，乘势而上，快马加鞭，推动“一带一路”建筑类大学国际联盟建设行稳致远，迈向更加美好的未来。

昨天晚上我们召开了预备会，与会专家对成立“一带一路”建筑类大学

国际联盟给予高度评价，特别感谢联盟成员的信任，同意我校担任“一带一路”建筑类大学国际联盟秘书长单位和首任轮值主席单位。借此机会，作为联盟秘书长和首任轮值主席，我真诚地承诺，北京建筑大学将积极发展与联盟成员的合作伙伴关系，竭尽全力搭建开放共享的交流合作平台，为“一带一路”建筑类大学国际联盟的顺利推进贡献力量，为成员做好服务工作。

今天下午，各位代表还将紧密围绕“在‘一带一路’发展倡议下，开展建筑类大学国际交流合作与创新人才培养”这一主题，进行深入的探讨和交流，希望大家畅所欲言、深入研讨、相互借鉴，为推进联盟的下一步工作和可持续发展，积极献计献策。

“会当凌绝顶，一览众山小”，我国唐代大诗人杜甫的诗句千古传诵，就是因为他表达了不怕困难、敢于攀登绝顶的雄心壮志，今天我们更要有“登月球而小地球”的宽广胸怀和远大抱负。

让我们把“一带一路”倡议变成行动，把“一带一路”愿景变成现实，为建设更加和谐宜居的地球、建设更加美好的世界培养更多的创新人才，做出我们更大的贡献，共同铸就“一带一路”发展的新丰碑！

谢谢大家！

Bring the Initiative into Actions, Transform the Vision into Reality, Innovatively Promote International Exchanges and Cooperation Among the Architectural Universities along the Belt and Road

Zhang Ailin, President of Beijing University
of Civil Engineering and Architecture

Honorable Mr. Ge Juzhong, Deputy Director of Beijing Municipal Education Commission, Distinguished Presidents, Dear Guests, Ladies and Gentlemen,

Today, we gather at Beijing University of Civil Engineering and Architecture (BUCEA) to witness the establishment of the Belt and Road Architectural University International Consortium (BRAUIC), and to discuss the innovative promotion of international exchanges and cooperation and innovative talent cultivation among the architectural universities alongthe Belt and Road. On this occasion, on behalf of BUCEA and all founding members of the BRAUIC, let me express my heartfelt thanks to you for your active support for the founding of the BRAUIC and my warm welcome to all guests present at today's Inauguration Ceremony and University Presidents'Forum.

In September and October 2013, Chinese President Xi Jinping, during his visit to Central Asia and Southeast Asia, launched the Belt and Road Initiative of jointly building the Silk Road Economic Belt and the 21st Century Maritime Silk Road. In this May, the remarkable Belt and Road Forum for International Cooperation was successfully held in Beijing, during which President Xi Jinping proposed to

build the Belt and Road into a road for peace, a road of prosperity, a road of opening up, a road of innovation and a road connecting different civilizations. The participating parties sent out a positive signal to jointly push forward regional and international cooperation, and to build a community of shared future for mankind. This forum has further clarified the cooperation prospects for the Belt and Road countries, planned a specific road map for the Belt and Road development, and launched a number of key projects for the implementation of the Initiative.

The Belt and Road Initiative has three connotations: the highlight of the characteristics of connectivity in the 21st century, the establishment of the economic development belt, and the prevalence of mutual benefit and win-win results. Featuring peace and cooperation, openness and inclusiveness, as well as mutual learning and mutual benefit, the Initiative brings enormous co-development opportunities and areas for the countries and regions along the Belt and Road.

As we have often said, a nation will prosper only when its young people thrive. Mutual learning and exchanges among young people as well as the exploration and innovation of young people are of utmost importance to the implementation of the Initiative and the sustainable development of the Belt and Road countries. To enable the next generation of young people in the Internet era to fulfill their dreams, we shall build up more platforms and open up more channels for cooperation, promoting the substantive international cooperation in higher education, and improving the level of higher education cooperation.

As the one and only architectural university in Beijing, BUCEA is jointly sponsored by the Ministry of Housing and Urban-Rural Development of China and the People's Government of Beijing Municipality. With a long history in architecture-related disciplines, the university, founded in 1907, has embraced its 110th anniversary this year. At present, we have established long-term cooperative relationships with 63 universities from 28 countries and regions. In 1999, we set up the Sino-French Training Center for Energy-related Professions together with the French government. Now, we have 26 high-end scientific research platforms at provincial and ministerial levels, including Beijing Advanced Innovation Center for Future Urban Design, and Engineering Research Center on Representative Ancient Building and Ancient Building Database of the Ministry of Education. What's worth mentioning is the university's commitment to architectural heritage protection and conservation along the Silk Road. We have successfully completed some world-renowned conservation projects, including the world's largest linear heritage corridor project—ancient temple relics in Xinjiang area of the Silk Road, the Great Wall, the Yungang Grottoes, and the Chau Say Tevoda in Cambodia and another world cultural heritage conservation projects.

In the last two years, through developing new partner colleges and universities and signing cooperation agreements, we have carried out extensive cooperation with colleges and universities from the Belt and Road countries, witnessing a new chapter in exchanges and cooperation featuring faculty and student exchanges, joint scientific research, and collaboratively-held international conferences. In addition, we have hosted the Beijing International Summit on Urban Design, the International Symposium on Contemporary Problems of Architecture and Construction, the Asia-Pacific Urban Construction and Management Practice Forum, the International Symposium on Spatial Data Infrastructure Construction and Application, Historical Buildings along the Belt and Road: Architectural Photography & Hand-drawn Art Exhibition, and so on.

In 2016, the university was granted the Belt and Road Scholarship for foreign students by the Beijing Municipal Education Commission, which was set to sponsor ambitious young people from the countries along the Belt and Road to study in Beijing. In September this year, the university was listed among the first batch of Belt and Road National Talent Cultivation Base with special funds for international architects and civil engineers cultivation and relevant discipline development. We have invited a great number of renowned international experts to give lectures at the university and dispatched several faculty delegations to go abroad for capability enhancement, thus having improved the scientific research cooperation level and the faculty members'overall ability. All these efforts have laid a solid foundation for the establishment of the BRAUIC.

Facing with four significant development opportunities—the accelerating pace of infrastructure construction along the Belt and Road, China's construction industry transforming and upgrading toward informatization, greenization and industrialization, the capital Beijing's new positioning and its goal of becoming a world-class harmonious and livable city, and the coordinated development in Beijing, Tianjin and Hebei regions and the construction of Xiong'an New Area, BUCEA has devoted to building itself into an open and innovative architectural university with world-wide recognition, pushing forward exchanges and cooperation with universities from all over the world, promoting comprehensive education reform and globalization, and to setting up an open and shared platform for international talent cultivation, scientific research collaboration and innovation as well as people-to-people and cultural exchanges. To this end, we came up with the idea to establish the BRAUIC which was immediately recognized and supported by the Beijing Municipal Education Commission and the Beijing Municipal Foreign Affairs Office. After sending out the written Proposal and the drafted Charter, we have met up with quick and positive responses and strong support from universities both at home and abroad. Since this May, we have received confirmation letters from 44 universities of 19 countries, including Russia, Poland, France, the United States, the United Kingdom, Armenia, Bulgaria, Czech Republic, South Korea, Malaysia, Greece, Nepal and Israel, to become a founding member of the BRAUIC. Today, 51 delegates representing 25 universities from 11 countries are present at the Ceremony to witness the formal establishment of the BRAUIC. We believe, with its further development, more and more universities from China and other countries along the Belt and Road will join the BRAUIC.

In order to promote the sustainable development of the BRAUIC, I would like to propose the following four suggestions:

Firstly, we should establish a management and operation mechanism featuring sustainable development for the BRAUIC, and set up an Open-End Fund.

Secondly, we should push forward the opening up and sharing of member universities'resources by means of informatization.

Thirdly, we should set up an Open-End Fund to support members for collaborative innovation and research in terms of talent cultivation, scientific research, faculty member development, campus culture, social services and other aspects, therefore promoting the integration of different advantageous and characteristic disciplines of different universities.

Fourthly, we should establish a landing mechanism for the achievements and proceedings at the university presidents'forum of the BRAUIC.

There are two ancient Chinese sayings. One is, "Well begun is half done" . Today marks the formal founding of the BRAUIC and a solid first step in its development. Such a good beginning greatly boosts our confidence. The other goes, "The last leg of a journey just marks the halfway point" . We still have a long way to go and arduous missions to fulfill. We must exert the utmost effort, ride on the momentum and accelerate the speed to promote the stable development of the BRAUIC for a brighter future.

Last night we held a preparatory meeting, and the participating experts gave high praises on the establishment of the BRAUIC. We are particularly grateful to the members of BRAUIC for their trust and recognition for our university to serve as the secretary-general unit and the first rotating presidency holder of the BRAUIC. As the secretary-general and the first rotating president of BRAUIC, I would like to make a solemn commitment that BUCEA will actively develop partnerships with BRAUIC members, and make every effort to build an open and shared platform for exchanges and cooperation, make our contributions to the smooth progress of BRAUIC and provide sound service to the members.

In the coming Forum under the theme of "International Exchange & Cooperation and Innovative Talent Development for Architectural Universities under the Belt & Road Initiative" this afternoon, I hope that by engaging in full exchanges of

views, we will contribute to pursuing the sustainable development of the BRAUIC in the future.

"When shall I reach the top and hold all mountains in a single glance" is a verse of a poem by Du Fu, a prominent Chinese poet of the Tang Dynasty. It has been read through all ages for its expression of the great ambition to conquer the difficulties and to climb up to the top without any fear, inspiring us to maintain a broad mind and great ambitions.

Let's bring the Belt and Road Initiative into actions, and transform the Belt and Road vision into reality! Let's make greater contributions to more harmonious and livable earth and a better world through cultivating more creative talents! Let's build a new monument together for the Belt and Road development!

Thank you very much!

“一带一路”倡议：未来城市发展与创新能力发展

亚美尼亚国立建筑大学校长　加吉克·加斯蒂安

尊敬的各位建筑类大学、科技类大学的同仁们、校长、副校长、教授及各位代表：

非常荣幸有机会参加“一带一路”建筑类大学国际联盟成立大会暨校长论坛，在此我向各位致以诚挚的敬意。

中国政府提出的“一带一路”倡议，旨在增进和平合作、开放包容、教育交流与互利合作。

“一带一路”倡议的落实，并不局限于意识形态、发展道路、发展模式、发展水平或其他标准，其目的在于在政治、贸易、基础设施发展、文化交流等领域谋求共同利益。

“一带一路”倡议已成为一种全球化模式。这种模式依托于开放的经济政策、贸易和经济平台，为加深各国在民间、文化、教育、科技、经贸等领域的合作关系创造了更多的机会。

我们赞同倡导者对这一概念的信念，认为迫切需要发展各国在双赢互利原则基础上的合作范围，并赋予其生命力。

宏伟的“一带一路”倡议必将促进亚欧非三大洲的经济发展。目前，“一带一路”倡议是规模最大、包容性最强的倡议。可以信心满满地宣布，落实这一倡议，就是要在六个主要目的地——即所谓的“走廊”上，建设桥梁、铁路、隧道、工程结构和旅游景点等基础设施。

显然，这一倡议将推动参与这一宏伟倡议的国家及其近邻的经济发展，激励金融机构，大幅增加需求；特别是建筑、建设和工程服务的需求。根据未来城市发展需求和创新能力发展的要求储备专业人才，是该领域高等教育机构面临的一项挑战。

女士们，先生们，亲爱的同仁们，

但是，城市发展的未来是什么样子的？谁将是未来城镇、道路和桥梁的设计者和建筑者？他们需要具备哪些知识和能力才能应对不断发展变化的世界提出的要求？

我们强烈认为，符合未来城市发展愿景的专业人才教育面临的挑战，需要有以下主要目标：

首先，教育作为传播知识的一种手段，应该成为创新能力发展的技能形成手段，成为理解客体和现象本质的方式。这绝非易事。我们必须努力实现这些目标，同时牢记我们数个世纪以来所积淀的经验、知识和建筑传统。

其次，是开展研究，以及传授以人为本（以人为中心）的、安全的、可持续城市和基础设施的设计与建设知识。未来的城镇应该在自然灾害和人为灾害面前得到更加充分的保护，应该是安全舒适的居住场所。很难用今天的眼光简要地描述未来城镇的样子。未来的城镇既可能是巨型城市，以交通线路为中心，各种垂直建筑鳞次栉比，并通过高架连接；也可能是航空城，围绕独立机场或古典城镇园林等建成。然而，无论设计特点如何，这些未来城镇都应该为国民创造稳定的环境，使其免受各种风险，并且生活期间不会浪费额外时间。这是一种传播健康、自给自足的生活方式，自由流动和科学创新的典范城市。然而，这种以纯粹的初级本地化典范为形式的环境，除非把人的因素考虑进来，否则不可能长期持续。后者在人类的进化过程中，对人的要求几乎没有任何改变，人类只有在一个激发内心、让内心充满惊奇和激动的环境中才能生存。

再次，研究与传播旨在创建资源节约型、环境友好型城镇的知识，未来的城镇和基础设施在设计、建造、经营和拆除阶段应与环境更加协调一致，不会对自然环境造成不可逆转的影响。考虑到全球迅猛的城市化趋势，我们认为把“绿色”思想引入城市发展非常重要。联合国“里约＋20”峰会和“人居三”会议框架下的城市评估正是在此背景下出现的。

最后，研究并传播知识，创建智慧城市。“智慧家园、智慧社区、智慧城镇，互联建筑和物联网”，都是听起来很超前的想法。将信息技术引入城市规划和建筑领域，极大地促进了设计过程、信息交流和建筑项目的组织过程。然而，这仅仅是这些领域结合给我们带来好处的开始。城市规划数字化，也就是建筑信息建模（BIM）广泛而深入的发展，将导致出现新一代的城镇规划。“智慧城镇”概念，蕴含现代化技术实施机会，追求的目标是为人类生活、工作、娱乐创造有利和安全的生活空间，自然环境和人造环境协调组合、相互补充。

因此，我们大学的目标是为未来的城镇和基础设施培养设计师、建筑师、决策者和公务员。

女士们，先生们，亲爱的同仁们，

我提出如下具体建议：

建议一：九年前，北京建筑大学和亚美尼亚建筑设计大学共同发起了名为“建筑与建造当代问题”的国际会议，致力于探讨本领域当前出现的问题。随后，切斯托科沃科技大学、圣彼得堡国立建筑大学、俄斯特拉发技术大学、布拉格捷克理工大学、佛罗伦萨大学和巴统绍塔·鲁斯塔维里国立大学加入共同主办。参会者涵盖的地理范围要大得多，不限于上述提及的大学。北京建筑大学承担十周年大会的组织工作，该会议将于 2018 年召开。

我要向我的合作伙伴尊敬的张爱林博士及其团队表示诚挚的感谢。

我建议在新成立的联盟之内扩大参会范围，使该会议成为建筑大学联盟各建筑大学之间沟通与合作的科学平台之一。各位尊敬的合作伙伴，我诚邀各位参加 2018 年十周年大会，践行联盟有关科学交流的重要论题。有了各位的参与，大会的地缘会进一步扩大，也许下一届会议——2019 年第 11 届会议，就会在贵校举行。

建议二：拟订振兴科教交流的议题。我们建议针对年轻一代启动暑期班。每年由一个参会国举办一期，给年轻一代创造接受杰出专家指导的机会，并让他们熟悉东道国的科教环境和文化。暑期班将揭示大学创新发展的技术和建筑建造专业。

尊敬的合作伙伴，作为落实本建议的第一步，我们愿意在我们大学，亚美尼亚国立建筑大学组织首期暑期班。我们建议讨论确定暑期班的使命。

建议三：在发展中国家，特别是某些地区，电信领域首先缺乏高水平的基础设施。在这方面，我们建议，根据我们大学的能力，利用现有的资源和积累的经验，并考虑到区域的重点任务，我们应该建立具有区域意义的研究和生产实验室，配备必要的现代化设备、最好的通信和电信手段（互联网连接、搜索能力和方案、信息数据库、适当的培训体系等）。

同时，区域研究和生产实验室成立及其成功开展活动的基本条件是，激发地方自治机构、商界，特别是中小企业的兴趣和积极参与，因为它们是重要的新增就业之源，它们能够解决好一大批社会经济问题。

建立此类研究中心，有助于有效、彻底地解决各种各样的问题，并通过共同努力，至少在特定地区内成为国家经济的推动力之一。

女士们，先生们，

尊敬的各位合作伙伴，

最后，我想再次向本次会议的中国发起人表示感谢。

伟大的孔子说过："知者不惑，仁者不忧，勇者不惧。"

我们十分希望，我们的努力不日就会对人们的日常生活产生积极的影响。

The Belt and Road Initiative: Future Urban Development and Innovative Capacity Development

Gagik Galstyan, Rector of National University of Architecture and Construction of Armenia

Dear colleagues, Presidents, Rectors, Vice-Rectors, professors and representatives of the Universities of Architecture and Construction, and Technical Universities,

I cordially greet you at the inaugural international consortium of the architectural universities and Presidents'Forum, organized within the Belt and Road Initiative.

The Government of the People's Republic of China has proposed a tremendous cooperative initiative to boost peace and collaboration, openness and tolerance, mutual educational exchange and benefits, within the framework of the Belt and Road Initiative.

The implementation of the Belt and Road Initiative is not restricted to the ideology, path, model and level of development or other standards; but is aimed at gaining mutual interest in politics, trade, infrastructure development, cultural exchange, and other spheres.

The Belt and Road Initiative is believed to become a globalization model, which assumes an open-door economic policy, trade and economic platform; with a deepening opportunity to develop and collaborate civil and cultural, educational, scientific, trade and economic ties between different countries and nations of the world.

We share the advocates'conviction of this concept who consider it urgent to develop and put into life scope of collaboration between countries based on a mutual interest in a win-win principle.

The mega Belt and Road Initiative will lead to economic progress on three continents-Asia, Europe, and Africa. Currently, there are no initiatives of such massive scale and inclusiveness, other than the Belt and Road. It is worthwhile to state with firm confidence, that putting this initiative into execution means to build infrastructures, such as roads, bridges, railways, tunnels, engineering structures, and tourist attractions with six main destinations-the so-called "corridors" .

Obviously, this initiative will give impetus to the economic development of the countries involved in this mega-project and their bordering neighbors, energize financial institutions, and sharply increase demand; especially for architectural, construction and engineering services. This is a challenge to the institutions of higher education in this sphere, to prepare specialists by future urban development demands and innovative capacity development.

Ladies and Gentlemen, dear colleagues,

But what is the future of urban development? Who will be the designers and builders of the future towns, roads, and bridges? What kind of knowledge and abilities will they need to be armed with to meet the requirements of a dynamically changing world?

We strongly believe that the challenges for education of specialists in complying with the vision of future urban development will need to have the following main goals:

First, education as a means of transferring knowledge, should be turned into a means of skill formation for innovative capacity development, and understanding the very essence of objects and phenomena. This is no easy task, and we must work hard to achieve these goals, at the same time keeping in mind the experience, knowledge and building traditions that our countries have held for many centuries.

Secondly, to carry out research and impart knowledge for the design and construction of human-focused (anthropocentric), safe, and sustainable cities and infrastructures. Towns of the future ought to be more protected from natural and anthropogenic disasters, and should be secure and comfortable settlements to live in. It's difficult to describe in short today's vision for the towns of the future. They can be both gigantic towns, that is to say "transpolir", centered around transport routes and densely built areas with vertical construction closely interwoven with each other by horizontal overhead connections; or "aerotropolis" organized around detached airports or small classical town-gardens, and so on. All of them, irrespective of their design characteristics, ought to create stable environments for their citizens, who are protected from risks, and live with no additional wasted time. This is a model propagating a healthy and self-sufficient life-style, free mobility, and scientific innovations. Nevertheless, this environment in the form of a purely elementary localized model, cannot be long-term sustainable, unless a human factor is taken into account. The latter has remained nearly unchanged in the requirements of a human being throughout all its evolution, who only survives in an environment which can kindle, surprise, and thrill them psychologically.

Thirdly, to do research and disseminate knowledge targeted at the creation of resource-saving towns, with the least impact on the environment. Towns and infrastructures of the future should be more harmonized with the environment in design-construction-operation and dismantling stages, without creating an irreversible impact on natural surroundings. Taking into consideration the impetuous urbanization tendency worldwide, we consider it important to introduce "green" ideology into urban development. Within this context is the urban evaluation of the UN "After Rio + 20" and "Habitat-III" framework.

Finally, to perform research and impart to create smart towns. "Smart home, smart community, smart town, connected buildings, and the internet of things" are ideas that will sound normal tomorrow. The introduction of information technology into the spheres of town planning and architecture has highly facilitated the process

of design, information exchange, and the organization of building projects. Yet, this is just the beginning of the benefits expected from the combination of these spheres. Town planning digitalization, that is, broad and thorough development of Building Information Modeling (BIM) will result in town planning of a new generation. The "Smart Town" concept with modern technological implementation opportunities, pursues the aim of forming favorable and secure living spaces for a human being to live, work, and recreate; as a result of the natural and man-made environments'harmonized combination mutually complementing one another.

Thus, the objective of our universities is to educate designers, builders, decision-makers and civil service employees for the towns and infrastructures of the future.

Ladies and Gentlemen, dear colleagues,

Let me present concrete suggestions:

Suggestion One: It's already been nine years since the University of Architecture and Construction of Beijing and the National University of Architecture and Construction of Armenia jointly established an international conference titled: Contemporary Problems of Architecture and Construction; devoted to the current problems within this sphere. Later, Chestokhovo Technological University, Saint-Petersburg State University of Architecture and Construction, Ostrava Technological University, Czech Technical University, Florence University and Shota Rustaveli Batumi University joined as co-organizers. The geography of the conference participants is much larger and is not restricted to the universities mentioned. Beijing University of Architecture and Construction itself has taken upon the responsibility for the organizational work of the 10th-anniversary conference, planned to be held in 2018.

I express my deep gratitude to my partner honorable Dr. Ailin Zhang and his team.

My suggestion is to enlarge the conference participation within the newly-established consortium, making it one of the scientific platforms of our consortium for the communication and collaboration between Architectural Universities. Honored partners, I invite you to become participants of the 2018 10th anniversary conference and carry out the most important theses of the consortium on scientific exchange. With your participation you will enlarge the geography of the conference and who knows, maybe the next-the 11th conference in 2019, will take place at your University.

Suggestion Two: Developing the topic of reviving scientific and educational exchanges. We suggest initiating a summer school for the younger generation. It will be held annually in one of the participant countries, giving the younger generation a brilliant opportunity to be trained by leading specialists, and become acquainted with the host country's scientific-educational environment and culture. The summer school will enable revealing of the techniques of innovative development for the universities and specialties of architecture and construction.

Respectable partners,

As for the first step to accomplishing our suggestions, we are willing to organize the first summer school in our university, the National University of Architecture and Construction of Armenia. We suggest defining the mission of the summer school during the discussions.

Suggestion Three. There is a lack of high-level infrastructures, first in the sphere of telecom-in developing countries and especially in their regions. In this respect, we suggest that, based on the capacities of our universities, using the available resources and accumulated experience, as well as taking into consideration the regional key tasks; we should create research and production laboratories of regional significance, furnished with the necessary modern equipment, the best means of communications and telecommunications (internet connections, search capabilities and programs, information databases, appropriate training systems and so on).

Meanwhile, the essential conditions for the foundation of regional research and production laboratories and their successful activities, are the interest and active participation of the local self-government institutions, and the business world, especially small and medium businesses which will become an important sources for new job creation, and able to solve a good number of social and economic problems.

Establishment of such research centers will contribute to the effective and radical solution to a great number of various problems, and through joint efforts, may become one of the moving forces for our national economies, at least within the given regions.

Ladies and gentlemen, dear Partners,

In conclusion, I would like once more to express my gratitude to the Chinese initiators of this event.

The great Confucius said: "Education breeds confidence. Confidence breeds hope. Hope breeds peace."

We fully hope that our efforts in no time will have a positive impact on our citizens' daily life.

在“一带一路”建筑类大学国际联盟成立大会上的讲话

沈阳建筑大学校长　石铁矛

尊敬的同事们、亲爱的朋友们，女士们、先生们：

大家上午好！

首先，请允许我代表沈阳建筑大学，以及联盟的共同发起单位，向参加“一带一路”建筑类大学国际联盟成立大会的各位来宾、各位朋友、各位专家，表示热烈的欢迎！更要向为了此次大会的顺利召开做了大量筹备工作、付出辛勤汗水的北京建筑大学的各位同仁，表示由衷的感谢！

我们都知道，“一带一路”是中国政府提出的一项意义重大、影响深远的国际合作倡议，它是在深入分析中国以及世界发展需求的基础上提出的，反映了中国和“一带一路”沿线国家和地区的共同愿景。

倡议实施近4年来，中国政府积极推进“一带一路”建设，加强与沿线国家和地区的沟通磋商，推动与沿线国家和地区的务实合作，实施了一系列政策措施。2017年5月14日至15日，首届“一带一路”国际合作高峰论坛在北京举行，这是“一带一路”倡议提出近4年来最高规格的论坛活动。高峰论坛明确了“一带一路”建设合作的目标和方向，确定了重点合作领域和行动路径，取得了一系列积极成果。

如今，国际社会和“一带一路”沿线各国和地区间已达成越来越多的共识，那就是“一带一路”沿线国家和地区在自然环境、社会政策、经济和文化等方面存在巨大差异，大家都面临着复杂的发展挑战。只有通过国际合作和创新发展，才能更好地应对各方面的挑战。

为了推动“一带一路”沿线国家和地区大学之间在教育、科技、文化、人才培养等领域开展全面的交流与合作，从2015年5月至今，新丝绸之路大学联盟、“‘一带一路’高校战略联盟”、“一带一路”国际教育发展大学教育联盟、“一带一路”中波大学联盟先后成立。这一个个“一带一路”大学联盟、高校联盟和教育联盟的成立，对于推动“一带一路”沿线国家和地区在校际交流、人才培养、科研合作、文化沟通、政策研究、智库建设等方面的交流与合作，培养具有国际视野的高素质、复合型人才，促进世界经济增长及欧亚地区的发展建设，具有重要的意义。

作为建筑类大学，我们有着相近的发展现状和特点，在“一带一路”建设发展上，我们有着天然的学科和专业优势，这是我们开展合作交流、实现互利共赢的坚实基础。通过成立联盟，我们能够共同探索建筑类人才跨文化培养与跨境流动的新型培养模式，促进联盟高校间师生互动，更好地致力于科技创新、人才培养、文化交流，促进“一带一路”沿线地区的共同发展。

多年来，沈阳建筑大学围绕“一带一路”倡议，在国际化人才培养和办学国际化等方面开展了一系列有益的探索。我们与罗马尼亚特来西瓦尼亚大学合作共建孔子学院，并多次成功举办“一带一路，人文交流”孔子学院夏令营活动；我们与波兰琴希托霍瓦工业大学共同申办孔子学院，国家孔子学院总部建议可以设立孔子课堂，逐步建设孔子学院；我们与俄罗斯顿河国立技术大学、德国维斯玛大学等40多个国家和地区的60多所大学建立了校际交流或科研、合作办学关系，我校的“中芬（北欧）木结构节能环保别墅建筑示范项目”和“中德节能示范中心”搭建了高端国际科技合作平台；我们新成立了交通工程学院和商学院，以国家“一带一路”建设大环境和东北地区经济发展为背景，为国家“一带一路”建设和东北地区经济建设培养高技术人才。

“一带一路”倡议的深入实施，对于国际工程人才培养和建设人才的培养提出了更大的需求和更高的要求，我们呼吁建筑类大学联合起来，共同探索建筑类学科及建筑类大学创新发展机制。我们成立“一带一路”建筑类大学国际联盟，就是要通过合作办学、专业共建、联合培养、合作创新等方式，推动“一带一路”建筑类大学的全面交流合作，推动教育综合改革和教育国际化进程。

“一带一路”倡议秉承的是“开放、包容、互利、合作”的原则，我愿意和大家一道，在此原则的引领下，秉承“和平合作、开放包容、互学互鉴、互利共赢”的发展理念，为“一带一路”建筑类大学未来的发展贡献自己的智慧和思想。同时，我也希望此次会议能够加深并拓宽我们在“一带一路”建设合作方面的立场与共识。最后，衷心祝愿大会取得圆满成功！

谢谢大家！

Speech at the Inauguration Conference of the Belt and Road Architectural University International Consortium

Shi Tiemao, President of Shenyang Jianzhu University

Honorable colleagues, dear friends, ladies and gentlemen,

Good morning!

First of all, please let me warmly welcome all guests, friends and experts who are participating in the Inauguration Conference of the Belt and Road Architectural University International Consortium on behalf of Shenyang Jianzhu University and the joint sponsors of the Consortium; furthermore, let me show heartfelt thanks to Beijing University of Civil Engineering and all architectural colleagues who have done much work for the successful holding of this conference.

As we all know that, the Belt and Road is a significant and profound international cooperation proposal proposed by the Chinese Government, and it was proposed based on deeply analyzing China and world development demands. It reflects the shared vision of the countries along the Belt and Road of China. Since 4 years from the implementation of the proposal, the Chinese Government has actively promoted the construction of the Belt and Road, strengthened the communication and negotiation with the countries along it, promoted the practical cooperation with these countries and implemented a series of policies and measures. On May 14th, 2019, the 1st The Belt and Road International Cooperation Summit Forum was held in Beijing, which was the forum activity with the highest specification since four years from the proposal of the Belt and Road. At the Summit Forum, the target and orientation of the construction and cooperation of the Belt and Road was indicated clearly, the key cooperative fields and action routes were confirmed and a series of active achievements were achieved.

Today, international society have reached more and more consensus with the countries and regions along the Belt and Road, i. e. : natural environment, social policies, economy and culture, etc. of these countries have certain differences and all of these countries are facing complex development challenges, so that, only through international cooperation and innovative development can they better respond to challenges of all aspects.

In order to promote the education, science and technology, culture and talent cultivation, etc. among the universities of the countries and regions along the Belt and Road, comprehensively carry out communication and cooperation, the University Consortium of new Silk Road, the Belt and Road Universities Strategic Consortium, the Belt and Road International Education Development University Education Consortium, The Belt and Road China-Poland University Consortium, etc. have been established from May 2015 till now. The establishment of all of these Belt and Road University Consortiums, University and Colleges Consortiums and Education Consortiums had communicated and worked for promoting the inter-academic communication, scientific research cooperation, cultural communication and policy research as well as think tank, etc. of the countries along the Belt and Road, and it has significant meaning for cultivating compound talents with international view and high quality and promoting world economic growth and development and construction of Eurasia.

As an architectural university, all of us are with the mission of cultivating talents for the construction of human living environment. On the aspects of construction and development of the Belt and Road, we have similar construction of subjects and disciplines, which is the shared basis for us to communicate and cooperate.

Today, all of us establish such a consortium with the hope of jointly exploring

to cultivate cross-bordering, trans-disciplinary and interdisciplinary new talents for urban and rural areas, promote interactions among teachers and students of the universities and colleges participate in the consortium, better devote to scientific innovation, talent cultivation, cultural communication and promote the overall economic development of the Belt and Road region.

Since many years, Shenyang Jianzhu University has made beneficial exploration on the aspects of internationalized talent cultivation and internationalization of school running surrounding the Belt and Road proposal. We cooperated with Romania Rerasy Manea University and established Confucius Institute, and have successfully held the Belt and Road summer camp activities for many times. We jointly applied and sponsored Confucius class with the Czestochowa University of Technology of Poland, we have obtained the first batch of Silk Road Chinese Government Scholarship Project, have established the foundation for talent cultivation, our university has established university-level communication relationship with over 80 universities in over 40 countries and regions, such as Russia Don State Technical University, Germany Wismar University, etc. In the operation process, we established Traffic College and Business College for the Belt and Road, which can cultivate more talents for the country. It shall be said that, Shenyang Jianzhu University is located in northeast China, and is in developing times. We will cultivate high technical talents for northeast region and construction of town and country planning. The further implementation of the proposal of the Belt and Road proposes higher requirements for engineering cultivation and construction talent. We seat here to discuss and agree with the suggestions jointly proposed by President Zhang Ailin just now. So, we express here to actively response and perform our joint responsibilities. We establish the Belt and Road architecture university consortium to promote the comprehensive development of all members of the Belt and Road architecture universities and promote the comprehensive reform of our education and the internationalization process of education through cooperative university operation, co-construction of disciplines and joint cultivation as well as cooperation innovation, etc.

Just now, the President of Armenia University also proposed good suggestions, and he has not only implemented the consortium and our purpose, but also expanded our communication and carried out more activities. The Belt and Road persists on the principle of openness, inclusiveness and mutual benefit cooperation. I am willing to be led by this principle together with you to persist on the development concept of peaceful cooperation, openness and inclusiveness, mutual learning and identification, mutual benefits and win-win, contribute our wisdom and thought for the future development of The Belt and Road architecture universities. And at the same time, I also hope this conference can deepen and expand our cooperative common view on the aspect of the Belt and Road.

Finally, congratulations on the success of the conference, thanks!

在“一带一路”建筑类大学国际联盟校长论坛上的报告

美国夏威夷太平洋大学校长　约翰·乔坦达

非常感谢刚才大家的鼓掌，我今天非常高兴能够来到北京，我也觉得非常荣幸，能够代表夏威夷太平洋大学，成为“一带一路”建筑类大学国际联盟的奠基成员之一。其实中国和夏威夷地区之间的关系，可以用特殊和源远流长来形容。事实上中国在夏威夷的足迹早在1789年就已出现，在此之前胡克船长曾抵达夏威夷，夏威夷被中国人称为檀香山。夏威夷人以热情好客著称，不管到哪里都会带上礼物。今天晚上我会向大家献上我的礼物，就是相思树做成的木舟，这代表我们旅途上的交通工具。在古老的夏威夷，有不少的探索者都是划着独木舟跨过太平洋，因此，木舟成为勇敢和发现的象征。

在夏威夷太平洋大学，我们有5000多名学生，有很好的多元性，他们来自世界上70多个国家和地区，我们在发现之旅上共同努力。我们理念非常简单，那就是鼓励学生终身学习，并且给学生新鲜的视角、新鲜的观念、新鲜的理念，这将使他们终身受益。我们有创新为本的项目，而且都是以市场为中心，以学生为本。我们学校的规模，一方面可以为学生提供选择，另一方面针对学生情况提供个性化的体验，我们有50多个本科生的项目、30个研究生的项目供学生选择。我们有五大学院，分别是商学院、拓展与跨学科教育学院、卫生健康与社会学学院、博雅教育学院和自然与计算机科学学院。我们最受欢迎的学科主要包括国际商务、海洋生物学和护理，我们最新的工程项目也受到社会各界和学生的关注和支持。我们课程设置都是小班教学，每个班规模不会超过20人，学校师生比是比1∶13。我们以人为本的理念，不仅体现在小班教学，也体现在让学生的个性得到发挥，这就是说，要针对每一个学生的个人特色进行课程的设置，让他们有适合自己学习的体验，保障他们通过教育能够真正增强自己的能力，有利于自己的职业发展，并且满足市场的需求。我们的学习氛围非常独特，因为我们的学校坐落在太平洋沿岸，在这里见证了世界上快速的经济发展。在这样的环境之中，我们也看到东西合璧的氛围，我们一方面给学生提供整体化学习的场地，另一方面，无论课堂内还是课堂外都有实践的机会。我们坐落在檀香山，我们有各种各样的实习机会和校外导师的项目，我们的学生可以直接在课堂之外和实习中锻炼自己的技能，在他们毕业的时候，可以更好地直接对接市场。

我跟大家稍微介绍一下我们非常尖端的科学研究项目，在我们的海洋研究所中，不少的项目正在进行。我们的研究者正在致力于科研项目，主要是为了提高人工养虾的抗病性。世界上人工养虾中60%的基因都可以溯源到我们海洋研究所。我们研究所是世界上首个人工养育珊瑚鱼的地方，比如说黄塘鱼，这有利于海洋和生态可持续发展。古代夏威夷人，靠一个人划桨，独木舟很难前行，每一个独木舟都必须有一个团队齐心协力来推动独木舟的前行，而这个桨就代表我们所有的教师人员和学生的通力合作。后来我越来越了解到丝绸之路的精神，概括起来就是和平合作开放包容、互鉴互学合作共赢，我认为它和我们夏威夷的一句古语非常类似，中文的意思就是，“通过合作的话，我们可以一路前行”。那么我们与联盟当中各大高校进行联手，可以实现近期的资源互通。我也相信，我们可以构建合力，为学生提供更好的机会。我们学校是在充分利用自己独特的岛屿文化和丰富的文化氛围的基础上设置课程，让学生能够有更多丰富的想法，而且我们也希望我们能够真正地拓展夏威夷的精神，拓展夏威夷的文化，将这种文化在整个联盟当中进行传播和传承。我们也都知道，我们学生进入的市场将是全球化的市场，竞争非常激烈，而且是不断变化当中的市场，联盟通力合作，相信能够给我们的学生提供一个保护的机会，让他们在更好的氛围当中进行合作，发展出自己的国际视角，并且有一个跨文化的实践能力。作为教育人员，我们的责任就是，帮助我们的学生满足21世纪市场提出的要求。我可以有信心地说，学生之间跨文化的交流、教师之间跨文化的交流，也可以丰富我们的学习体验，丰富我们的项目课程。大家可以想象一下，一个跨境的虚拟的课堂看起来是什么样子？在这个课堂当中，来自亚洲、非洲和夏威夷的学生，坐在一起相互学习，并且从世界级导师那里进行学习。今天我们的技术已经成熟，这个愿景已成为现实，现在已在各个高校之间建立合作关系，“万事俱备，只欠东风”。

最后想说一下，我们这个木舟是由寇阿相思树作为材质，它代表夏威夷的精神，坚韧不拔，而我们的联盟也是非常强大，坚韧不拔，因为它是由世界知名的高校组成。我们有着共同的愿景，而且有着共享的资源，相信通过合作，我们会取得更大的成功。我们非常荣幸能够成为联盟的组成部分，而且我也是非常期待在未来可以和大家一起划桨，共建成功。非常感谢。

Report at the President Forum of the Belt and Road Architectural University International Consortium

John Gotanda, President of Hawai'i
Pacific University, the United States

Thank you all for your applauses. It is my pleasure and my honor to be in Beijing today, to represent Hawai'i Pacific University and become one of the foundation member countries of the Belt and Road Architectural University International Consortium. Actually, the relationship between China and Hawaii can be described as special, with a long history. As early as 1789, China has left its mark on Hawai'i. Before then, Captain Hu Ke once arrived in Hawai'i, which was called as Honolulu by Chinese. Hawai'ians are famous for their hospitality, and wherever they go, they will take some gifts. Tonight, I will also present my gift, a carpenter made of acacia rachis, which represents the traffic tool on our journey. In ancient Hawai'i, there were many explorers, and all of them crossed the Pacific Ocean by canoe. Therefore, carpenter becomes the symbol of courage and discovery.

In Hawai'i Pacific University, there are more than 5000 students from more than 50 states and more than 70 countries and regions, with good pluralism. I wish we can make efforts together on the journey of the discovery. Our idea is very simple; that is to encourage students to study all the time, endow them with new angles, a new concept, new ideas, which will benefit them lifelong. We have projects based on innovation, with the market as center and students as the basis. Our

school can offer different options for students, also provides an individualized experience for students. There are more than 50 undergraduate projects are available for option, and more than 30 postgraduate projects. Our university consists of 5 colleges, and they are College of Business, College of Expansion and Interdisciplinary Education, College of Health and Society, College of Liberal Arts, and College of Natural and Computational Sciences. Our most popular disciplines include international business, marine biology and nursing. Our latest engineering project also attracts the attention and support from different circles of society and students. Our curriculum setting is small class teaching, with less than 20 students each class. The teacher-student ratio in our school is 1∶13. Our idea of taking people as the basis not only refers to small class teaching, but also enabling the students to play their personality well. That is to say the curriculum is set according to characteristics of each student, making them to have the study experience suitable to themselves, to ensure that they can strengthen their ability construction, benefit for their vocation development and meet the needs of the markets through education. The study atmosphere of our students is unique, because our school locates on the coast of the Pacific Ocean, which witnesses the fastest economy development in the world. In such an environment, we have such an atmosphere of combining the west and the east, which not only offers the place for overall study for students but also practice opportunities either in the classroom or outside classroom. Located at Honolulu, we own various kinds of internship opportunities and projects of off-campus supervisors, and our students can practice their skills outside the classroom and in the internship directly. That is to say they can adapt to the market better when they get to graduate.

I want to introduce our cutting-edge scientific research project a little bit, especially some projects which are undergoing in our marine institution. The scientific research projects our researchers are engaged in are to improve the disease resistance of shrimp culture. 60% of the genes of the cultured shrimps are originated from our marine institution which is also the first place for coral fish culture. For example, yellow fish is beneficial for the sustainable development of the ocean and ecology. It is hard for the ancient Hawai'ians to row the canoe ahead alone, and each canoe should have a team to paddle together to sail ahead. And this paddle just represents the cooperation of all the teachers and students. Later, I understand the spirit of the Silk Road, which can be summarized as peace, cooperation, opening, tolerance, mutual learning, and mutual benefit. I think it is very similar to one old saying in Hawai'i, with its Chinese meaning as we can go ahead through cooperation. In this way, through the cooperation of various universities in our Consortium, we can realize the mutual communication of recent resource sharing. I believe, in such a way, we can also construct joint power for all the universities and provide better opportunities for the students. Our school sets the curriculum by using the unique island culture and rich cultural atmosphere, enable the student to explore more colorful ideas, and we also sincerely wish that we can really expand the Hawai'i spirit and Hawai'i culture, and spread and inherit this culture in the whole Consortium. As we all know, our students will enter into a global market, full of fierce competition and changes all the time. As a Consortium, we should be able to provide a protective opportunity for our students through cooperation, make them cooperate in a better atmosphere, develop their international angle and have the cross-culture practice ability. As educators, our responsibility is to help our students to meet the requirements posed by the 21st century. We are very confident to say that the cross-culture exchange among students and the cross-culture exchange among teachers can also enrich the study experience among universities, enrich the project curriculum. Can you imagine what a virtual classroom will look like? In this classroom, students from Asia, Africa and Hawai'i, will sit together and study from the world-class supervisor. Today, our technology has been mature, and this vision has become true. Now universities have established cooperation. Everything is ready, and all that we need is an east wind, a crucial action.

Finally, this wood paddle is made of Koa, which represents the firm and indomitable Hawai'i spirit, while our Consortium is also very powerful, firm and indomitable, because it is composed by the world-famous universities, with the same vision, shared resources. I believe we will achieve bigger success through cooperation and we feel so honored to become one member of the consortium, and look forward to paddling together with all of you in the future, to achieve success together. Thanks very much.

把准未来导向，创新合作机制，培养建筑类创新人才

北京建筑大学校长　张爱林

尊敬的各位校长，各位同事，按照原来的计划，首先应由俄罗斯建筑科学院副院长做报告，但是因为航班调整，他现在还没到，所以我变成了第一个发言。中国有一句话，叫“抛砖引玉”，第一个发言都是“抛砖引玉”，我先抛个“砖”，引出接下来演讲者的“玉”。我的报告题目是“把准未来导向，创新合作机制，培养建筑类创新人才”，报告内容主要包括三个方面。

第一部分，就是北京建筑大学概况。在给大家发的材料当中，已经介绍了我校的基本情况。北京建筑大学始建于1907年，那时叫京师初等工业学堂，当时就有木工科，就是土木和木结构建筑专业。2017年是建校110周年，我校建筑类专业同样历史悠久，详细过程我就不细讲了。

与今天报告相关的，还有我们重要的创新，除了我们从本科生、硕士研究生、博士研究生到博士后流动站的全过程人才培养体系以外，2016年学校还获批“北京未来城市设计高精尖创新中心”。另外，我们跟国家文物局合作的“建筑遗产保护理论与技术”服务国家特殊需求博士人才培养项目也取得了不错的成绩。

从新中国“十大建筑”，到1990年的北京亚运会工程，再到2008年的北京奥运会工程，我校师生和校友都做出了重大的贡献。20世纪50年代，我校教授和老师也为北京市建设做出了重要贡献，北京市规划展览馆中就有我校校友赵冬日（1949—1958年任教）与朱兆雪提出的北京城规划方案，跟

梁思成先生的北京城市规划方案挂在一起。他们的方案虽有不同，但指导思想都是把北京古都保护好，在古城外面规划建设一个新城。非常遗憾，他们的方案没有实施。

我们学校有两大特色，一是从区域定位上长期服务首都北京的城市建设，就是北京味十足；二是从行业定位上服务国家建筑业转型升级，就是建筑味十足。去年，我校“未来城市设计高精尖创新中心”获批“北京高等学校高精尖创新中心”。作为科技创新特区，以五年为一周期，北京市每年给予5000万至1亿元的经费投入，原则上不低于70%的经费要用于聘任国内外高端人才，不低于50%的经费要用于引进国际顶尖创新人才。我校高精尖中心有五大研究方向，包括文化遗产保护与城市有机更新、绿色城市与绿色建筑、城市地下基础设施更新与海绵城市建设、城市设计与管理大数据支撑技术以及城市设计理论方法体系。

借助这样一个创新平台，我们创新科研组织运行模式，实行PI（学术带头人）负责制和组阁制，首先由首席科学家组成大学科团队，称为“一级PI”，下面是学术带头人组成PI学术团队，称为“二级PI”。我们已经聘请了国内外著名专家加入高精尖中心学术委员会和国际咨询委员会，比如荷兰代尔伏特大学的马克教授、美国哈佛大学的科克武德教授、宾夕法尼亚大学的迈克尔教授等。

我们服务国家和北京的重大工程包括：北京通州副中心旧城城市设计与更新、张家口崇礼2022年冬奥会越野滑雪场长城保护规划、故宫古建筑群数字化及三维重建、国家体育场（鸟巢）精密安装测量与安全监测、北京海绵城市建设试点项目、中荷未来污水处理技术研发中心、北京城市副中心综合管廊工程、北京大兴国际机场新型大跨度钢结构工程等。

经过深入研究，我校中长期发展规划确定学校的发展目标就是要建设“国内一流、国际知名、具有鲜明建筑特色的高水平、开放式、创新型大学”，这和“一带一路”倡议理念是相通的。

我要讲的第二部分，就是把准建筑类大学发展的未来导向和问题导向。为什么要把准未来的导向？我们今天教的学生是为未来储备的人才，他们是现在的学生，学校也是现在的大学，那么，我们应该教什么？学生应该学什么？我们应该怎么教学、怎么管理？我认为，应该努力为学生构造一个知识、能力、素质三位一体的培养方案，让他们能够满足未来的需求，既包括学生作为个体的未来，也包括世界、国家、教育、行业和区域的未来。他们在今天学到了什么、会做什么，将来才能学有所用？所以我们的教育是为未来储备人才，而不是我们今天教什么他们学什么，今天马上就用什么、会什么。这就是我们遇到的挑战：可能学生一毕业时，他们掌握的知识就成了过时的知识，而他们对新产生的知识怎么获取？

关于全球未来教育发展，大家都知道，联合国教科文组织第38届大会发布了《教育2030行动框架》。它的核心是迈向全纳、公平、有质量的教育和全民终身学习。它提倡开放的教育、可持续的教育、平等的教育、人本的教育和适合的教育。这是全球未来教育的蓝图。未来教育还有一个重要特点，就是信息技术革命推动教育革命和全球化，信息技术学习环境正建构着趋于开放式、连接式的教学模式，教师的角色从讲台上的讲解者，变成了引领者、示范者。在座有来自中国、日本、韩国的参会代表，围棋高手众多。世界围棋高手李世石几次被人工智能机器人阿尔法狗（AlphaGo）击败，赛后他对记者谈感受，他说他要让他的女儿和阿尔法狗（AlphaGo）学习下围棋，而不是跟他学下围棋，因为阿尔法狗（AlphaGo）的计算能力和速度远远超过他，他这个老师在某种程度上失去了作用，机器战胜了他。这就是信息革命带来的挑战。

2000年，中国的人均国民生产总值约1000美元，人民生活达到总体小康水平；2020年，中国要全面建成小康社会，人均国民生产总值近10000美元；2035年，我们要进入高收入国家门槛；2050年，要建成现代化强国，达到中等发达国家水平的建设目标。这就是中国的未来，这样的未来对于未来学生培养、对于建筑业发展的需求是什么呢？今天上午，我们已经讲到了“一带一路”倡议，今天下午各位专家还会进一步阐述“一带一路”的概念。一句话，我们要在全球化的视野下考虑我们的问题。

关于中国的城市化进程，1949年中华人民共和国成立的时候，我国城市化率仅有10.64%；经过漫长而曲折的城市发展时期，到1981年才增加到20.16%；改革开放多年以后，城市化率在2016年达到了57.35%，预计这一

百分比在2020年将达到60%左右，到2030年将达到70%左右。虽然跟世界发达国家80%左右的城市化率相比还有很大差距，但是对于我们而言是巨大的城市化发展进步。

为了实现这个目标，中国建筑业必须转型升级，转变发展方式。我们要向绿色化、工业化、信息化转型升级，2020年新建建筑50%达到绿色建筑要求。我们还要以推广装配式建筑为重点，大力推动建造方式创新。装配式建筑比例要达到30%，钢结构比例要达到10%。

另外，首都北京刚刚通过了新的总体规划，即《北京城市总体规划（2016—2035年）》（下文简称《规划》），其中包括北京城市副中心和河北雄安新区建设。《规划》提出，北京2020年发展目标是建设国际一流的和谐宜居之都实现阶段性目标。什么叫阶段性目标？就是“大城市病”等突出问题得到缓解，初步形成京津冀协同发展、互利共赢的新局面。2030年，北京基本建成国际一流的和谐宜居之都，治理“大城市病”取得显著成效，首都核心功能更加优化，京津冀区域一体化格局基本形成。到2050年，全面建成国际一流的和谐宜居之都，京津冀区域实现高水平协同发展，建成以首都为核心的世界级城市群。

以上就是我讲的“未来导向”，下面我们要进一步把准问题导向。问题很多，我主要归纳以下几条：第一，我们世界眼光、国际标准意识还不够，一方面是中国的教授和学生“走出去”比较少，另一方面我们将国际专家请进来，这次联盟校长论坛就是“请进来”，未来我们还要持续请更多国际知名专家学者，经常请你们到中国来，到北京建筑大学来。第二，就是我们的学科专业划分过细，因此割裂了多学科的有机融合。我举个例子，比如说智慧城市，你说是计算机专业研究智慧城市，还是信息控制专业、环境专业、建筑专业、土木专业的人研究智慧城市？其实，智慧城市是个多学科交叉领域，靠单一学科、单一专业解决不了问题。现在我们提出一个概念，叫“新工科”，这个具体定义目前正在讨论，但是它就是要回归到过去的大学科和多学科的概念。第三，就是创新实践解决复杂问题能力培养方面的不足。大家都知道《华盛顿协议》，2016年中国也加入了《华盛顿协议》，就是要以学生为中心，培养他们通过理论分析解决复杂工程问题的能力，而且要将这个能力的培养贯穿学生成长的全过程。建筑类大学未来发展，也要讲未来导向、问题导向、目标导向，就是要建设创新型建筑大学、建设一流建筑类学科、培养具有国际视野和通晓国际规则的一流建筑类创新人才。这与我们“一带一路”倡议中互联互通的理念是相通的。

大家都知道，大学有四项最基本的职能：人才培养，科学研究，社会服务，文化传承。今天我们“一带一路”建筑类大学国际联盟的成立就是大学第五大基本职能——国际交流与合作的体现。实际上，国际交流与合作职能是贯穿在人才培养、科学研究、社会服务、文化传承所有方面，每个方面都需要国际交流与合作，都需要互联互通，都需要资源开放与共享。

那么，建筑类大学发展途径选择就是开放、创新、合作。开放才能实现共赢和资源共享，是我们实现提质转型升级的必由之路；只有创新，才能解决新问题，解决了老问题，又产生了新问题，所以要不断创新，不断解决新问题；合作，就是一加一大于二，一加一再加一大于三，甚至大于N，也就是多赢的理念。

第三部分，要落实“一带一路”倡议的理念，就要创新机制，培养建筑类创新人才。因此我们发起成立建筑类大学国际联盟，在“共建、共享、共赢”的建设原则指导下，深度合作，培养建筑类创新人才。我们要全方位推动联盟合作，互联互通。在座各位都是大学校长，是大学的管理者和领导者。在教育管理与服务方面，我们要推进发展战略和愿景对接，促进教育的理念和资源互通，共建合作办学及学分互认项目。创新型人才培养方面，我们的合作形式也有很多，比如暑期学校、工作营、交换学习、短期访学、海外实习等。还有教师能力提升与发展方面的合作，包括教师互派、长短期访学、学术交流研讨等。教师是最关键的，只有我们的教师有创新能力，我们的学生才能有创新能力；我们的教师有国际视野，我们的学生才能有国际视野；我们的校长有国际视野，我们的管理团队才能有国际视野。在科技创新与社会服务上，当然更需要共同开展国际合作项目研究，我们希望依托北京建筑大学北京未来城市设计高精尖创新中心平台，邀请在座各个大学的专家都到北京建筑大学当教授。最后就是文化桥梁与纽带，就像我们联盟这个LOGO一样，我们真正变成桥梁，变成纽带。

最后，我想再次重复我上午讲过的一句话，就是我们要把倡议变为我们的行动，把愿景变为我们的现实，用创新机制推进“一带一路”建筑类大学国际交流合作！我有信心，我们一定会成功！谢谢大家！

Hold the Future Orientation, Innovate the Cooperation Mechanism and Cultivate Innovative Talents in Architecture

Zhang Ailin, President of Beijing University of
Civil Engineering and Architecture

Honorable presidents and colleagues,

It will be the turn of the vice president of the Russian Academy of Architectural Sciences to make the report according to the original plan of the manual. However, due to the adjustment of his flight, he has not arrived yet and will not arrive until this afternoon. Therefore, I become the first speaker. There is a Chinese saying that the first speaker is to throw out a minnow to catch a whale. I will serve as a modest spur to induce you to come forward. My topic is to hold the future orientation, innovate the cooperation mechanism, and cultivate innovative talents in architecture.

I will talk about three aspects.

In the materials sent to you, we have already introduced the basic information about our school. Beijing University of Civil Engineering and Architecture founded in 1907 and has been 110 years this year. At that time, we had the major of civil and wooden structures. Therefore, our school has a relatively long history in the field of architecture. Of course, the development time is relatively long. I will not elaborate on this process.

Let's focus on today's topic. Apart from the whole process of personnel training system from undergraduate and postgraduate students to doctoral students to postdoctoral mobile stations, we won the Beijing Advanced Innovation Center for Future Urban Design last year. Also, our doctoral personnel training in architectural heritage protection has done well in cooperation with the State Administration of Cultural Heritage.

In the history of our school, we can see from the top ten buildings since the establishment of new China that the teachers, students and alumni of our school have made important contributions to the Beijing Asian Games in 1990 and the Beijing Olympic Games in 2008. In the 1950s, our professors and teachers, for example, Zhu Zhaoxue had made plan for Beijing city, and the plan is linked to Mr. Liang Sicheng's plan. Their plans were different, but their ideas were to protect the ancient capital and to build a new city outside. Unfortunately, their ideas were not realized.

These projects were all done by our alumni and our students. In history, our school has two major characteristics. One is that we have been serving the construction of the capital Beijing for a long time. The second outstanding characteristic is that we make efforts in the transformation and upgrading of our national construction industry. Last year, Beijing approved us to build a high-tech innovation center for Beijing's future urban design. This is a special zone for innovation and technology. It invests at least 50 million RMB per year in a five-year cycle. 70% of its funds are used to employ high-end talents at home and abroad, of which 50% is not less than international talents. We have five major research directions, such as cultural heritage protection and organic renewal of cities, green cities and green buildings, underground engineering, urban theoretical big data, including sponge cities, environment and other disciplines.

With the help of such an innovative platform, we have innovated the operation mode of scientific research organizations. First, according to the PI system, the first level consists of a team of chief experts, followed by an academic team of academic leaders. We have already hired famous experts from other schools at home and abroad to participate in our high-tech centers. For example, these are professors from Harvard University, Michigan University, and this is Professor Mark from Holland, including Professor Michel, etc. For the project that we serve the country′s major needs, I want to give a few examples: one is the protection plan for the old city of Beijing, which is building a sub-center. As we all know, Zhangjiakou and Beijing will jointly host the 2022 Winter Olympics. We have made all the protection plans for the Great Wall in Chongli area, including the surveying and mapping of the Forbidden City and the Bird′s Nest, and our new sponge city pilot project in Beijing, including the sewage treatment research and development centers in China and Ho-lland. We recently completed the utility tunnel project of the Beijing Sub-center, and we have just completed the experiment of the Beijing new airport. Our future goal is to build a high-level, open and innovative university with outstanding architectural characteristics that is first-class in China, internationally renowned, and is consistent with our concept of "the belt and road initiative".

Secondly, I want to talk about how to make sure of the future orientation and problems of the development of architecture universities.

Why should we make sure of the future orientation? As you can see, the students we teach today are modern students. Our current university is today′s university. So what do we teach him? What do students learn? The relationship between time and space is that through how we teach, how we learn and how we manage, we can build a trinity of knowledge, ability and quality for students so that they can satisfy the future. What have students learned today about their future as individuals, the future of the world, the country, education, industry and regional economy? What they have learned and how they can use it in the future. Therefore, our education is to reserve talents for the future. Such space-time logic is not what we teach today, what they learn today, what they will use immediately today, what they will use today, but what they will meet in the future. This is the challenge we have encountered. It is possible that when we just graduated, the knowledge we learned today was outdated, and the student did not learn the newly generated knowledge.

As we all know, the 38th UNESCO General Conference issued the 2030

Framework for Action on Education. Its core is inclusive, fair, quality education and lifelong learning for all. It has open education, sustainable education and equal education. This is our blueprint for future education. Therefore, an important feature, the information technology revolution, has pushed forward the globalization of our education revolution. We have adopted an open and connected education mode. So our learning model and teachers role have turned out to be the lecturers of the platform and become today's leaders and demonstrators. AlphaGo is a software plays Go. China. Masters of Go from Japan and South Korea, like world Go master Li Shishi lost several times to AlphaGo in chess. He said that his daughter played Go with AlphaGo instead of playing Go with him. AlphaGo is his daughter's teacher. To some extent, he lost his role and the machine defeated him. Because of my learning ability, learning from machines is better than learning from you. This is the challenge brought by the information revolution.

In 2000, China was a general well-off society in which GDP per capita was about USMYM 1000. By 2020, China will build a moderately prosperous society in all aspects in which GDP per capita is about USMYM 10,000. By 2035, China will enter the lowest threshold for high-income countries. By 2050, we will have basically achieved modernization and enter the average level of developed countries. Therefore, this is such a future for China. What is the demand for China's future training and construction industry in such a future? Therefore, we have already talked about the proposal of "the belt and road initiative" this morning, and experts will also talk about the concept of "the belt and road initiative" this afternoon. We should consider our problems in this international perspective, that is, from the perspective of globalization.

In the second process of urbanization in China, when the People's Republic of China was founded, our urbanization rate in China was only 10. 64%. For a long time, in the course of our poverty, our urban greening rate was very low. It was only 20% in 1981. However, over 40 years of reform and opening up, the rate has become this curve. Now we reached 57. 3% in 2016, we will reach about 60% in

2020, we will reach about 70% in 2030, however 70% compared with the 80% of the world's developed countries is still a big gap, but this is a huge urban change.

To achieve this goal, our Chinese construction industry must transform and upgrade and change its development mode. We are transforming and upgrading to greenery, industrialization and informatization. 50% of our new buildings will reach the green building standard by 2020. We will vigorously promote innovation in construction mode with emphasis on the promotion of fabricated buildings. In addition, we will achieve 30% in fabricated buildings. I am a professor of steel structure. We have done a great deal in steel structure. Steel structure is less than 5% now, and we will double to 10%. Also, the capital Beijing has just passed a new general plan. By 2030, Beijing's overall urban plan will include the construction of sub-centers and the construction of Xiong'an new area. Therefore, in 2020, we will achieve the phased goal of a world-class harmonious and livable city. The phased goal is that the management of our major cities will have significant effects. In 2030, we will basically build a harmonious and livable city and in 2050 we will fully build a harmonious and livable city.

So this is what I said just now, the future orientation, the following is the problem orientation, the problem orientation is a lot, and I will sum up a few. One is that our awareness of international standards of world vision is not enough, our Chinese professors and students that go aboard are relatively few, so we invite you to come here, today we invite you to come here, and we will continue to invite you to come to China for a long time and often, to Beijing University of Civil Engineering and Architecture. And the second is that our disciplines are so finely divided that split up the organic integration of disciplines, let me give an example. For example, smart cities, we all know smart cities. Whether computer major is engaged in smart cities or information control is engaged in building cities, or construction, civil engineering s. In fact, smart cities are multidisciplinary cross-fields, and a single subject and a single major cannot solve the problem. At present, a new engineering course is proposed, which is currently under discussion, but it will return to the concepts of the past major disciplines and multi-disciplines.

Third, the ability to innovate and practice is still insufficient, so everyone knows that Washington Accord, China joined Washington Accord last year, which is student-centered, training their ability to solve complex engineering problems through theoretical analysis, and this ability training should run through the whole process of student growth. In future of architectural universities, the future orientation, problem orientation and goal orientation, that is, first of all, universities should be innovative. We should break through the concept of universities in the past. Students should also learn the concepts and influences of other universities, including first-class disciplines, including the importance of cultivating an international perspective and understanding international rules, because our "the belt and road initiative" concept is the concept of connectivity.

As we all know, universities have four basic functions. Among these functions, today's Belt and Road Architectural University International Consortium is international exchange and cooperation. In fact, it runs through all aspects of talent cultivation, scientific research, social services and cultural heritage. International exchange and cooperation are needed to link it up, interconnection is needed, and resources are needed to be open and shared.

Then the choice of development path for architectural universities is openness, innovation and cooperation. Only through openness can realize win-win results. Only through resource sharing can we transform our system. Only through innovation can we solve new problems. We have solved old problems, and we have created new problems. Therefore, we must constantly innovate and solve new problems. Cooperation means that one plus one is greater than two, one plus one is greater than three, or even greater than N, which is the idea of winning all.

Fourth, in the morning, I said that the concept of "the belt and road initiative" should be implemented, mechanisms should be innovated, and innovative talents should be cultivated. Therefore, in this architectural university international consortium, we have something to discuss, our platforms and shared platforms should be jointly built, our achievements should be shared, and win-win results should be achieved. We should cooperate deeply to cultivate construction talents, and all-round and all-region alliances should be exchanged. All of us here are presidents and leaders of university. Our education management and service should push forward the docking of development strategies and visions. We should exchange ideas and resources of education, and then cooperate in running schools. There are many ways to cultivate innovative talents. Just now, when we took a break, we said that our summer school work camp should be promoted in many ways, like short-term mutual visits, short-term learning, which is rich and colorful, specialized, subject-specific, classified, including teacher exchange. Teachers are the most critical part, and only our teachers have innovative ability that our students can have innovation ability. Only our presidents have international vision that our management team can have international vision. Therefore, in the scientific and technological innovation, social services, we need more international research and cooperation projects, etc. We hope that with the platform of high-tech innovation in future urban design of Beijing, experts from your universities present here will be professors at Beijing University of Civil Engineering and Architecture, and I will pay you.

Finally, we need cultural bridges and ties, just like our logo, we really become bridges and ties. I will repeat this words that I said in the morning, we will turn the initiative of "the belt and road initiative" into our actions, turn the vision of "the belt and road initiative" into our reality, and innovate and promote the cooperation of the Belt and Road Architectural University International Consortium. I believe that we will succeed, thank you all!

新时代保加利亚建筑领域高等教育的发展

保加利亚建筑土木工程和大地测量
大学校长　伊万·马尔可夫

各位尊敬的校长，各位来宾：

大家下午好！今天非常荣幸能够代表我们学校在此次校长论坛上发言。我要特别感谢“一带一路”建筑类大学国际联盟的各个发起单位，我们非常荣幸能够加入这个国际联盟。桥梁和道路的重要性不言而喻，而我们这个联盟，就是为教师和学生的交流搭建桥梁。

我们学校成立于1941年，成立之初是一所高等技术学校。由于各方面的原因，如今保加利亚已经没有这种高等技术学校了。后来，随着法律法规的变化，我们学校更名为建筑土木工程和大地测量大学。现在给大家展示的是一些老照片，包括我们学校教学楼照片、教授合影，照片中一些教授还是我当年上学时的老师。

接下来介绍一下我们的学院和专业设置情况。建筑学院设有建筑学和城市规划两个专业方向，建筑学方向仅有硕士项目，城市规划方向包括本科项目和硕士项目。大地测量学院设有大地测量硕士项目、土地和地产管理本科项目和硕士项目。我们还有结构工程学院、水利工程学院和交通工程学院，设有结构工程、建筑工程管理、水利工程、水利供应和污水处理、水管理等专业。我本人是结构工程学院的教师。同时，我们的交通工程学院开设了交通工程硕士项目，共包含10个学期的教学，学生在完成前4个学期的学习之后，需进行专门的实践工程学习。此外，在德语语言方面，我们和维也纳理工大学有一个合作项目。我们共有27个博士学位项目。现在大家看到的是我们学校得到的各类学术认证，包括英国皇家建筑师学会的认证证书。

我们学校大力开展各类科学实践活动，不断拓展各类国际项目，参与各类欧洲教育计划和国际科研项目。在建校 75 周年庆典期间，学校组织了一系列的科研活动，保加利亚前总统也曾来我校莅临指导。从 1991 年至 2001 年，共开展 16 个“坦普斯计划”项目；与欧盟 60 所大学和中学共同签署了“伊拉斯谟计划”和“中欧大学交换”项目双边协议；获批欧盟科技框架计划、北约研究计划、联合国研究计划等资助项目。

现在大家看到的是我们位于索非亚城市中部的校区。这一校区仅用于教学活动，照片中是一座教学楼。

当然，我们也面临一系列新的挑战，包括欧盟长期存在的竞争、欧洲一体化和 21 世纪对高等教育发展提出的新要求等。但是从现在开始，在“一带一路”建筑类大学国际联盟新倡议下，我们可以携手应对挑战。

我们有明确的意愿和目标，希望未来能进一步扩大我们的招生范围，可以通过这次大会进一步开展合作。

感谢大家的聆听。

The Development of Higher Education in Bulgaria in the New Era

Ivan Markov, Rector of University of Architecture,
Civil Engineering and Geodesy, Bulgaria

Distinguished presidents, dear guests,

Good afternoon! It is my honor to stand here to address at this forum representing my university. At the beginning of my address, I want to say I also discussed with other guests while I was preparing for this address. Originally, I wanted to introduce Bulgaria to all of you. However, I changed my mind later. Anyway, I want to thank the initiating unit of this Consortium especially. I am also so honored to be able to join this International Consortium.

First, in my introduction, I want to share that railway and road are certainly very important. However, now what is more important is to establish a bridge among the students and teacher, which is also an honor for us. Our university was founded in 1941. At present, there is no such higher technical school any more. For some reasons, I have no method to explain in details here. Our law articles also made some change, and our university changed a new name. As shown in the picture, these are our university buildings, these are our professor. Maybe some of the pictures are relatively old and some of the professors are still the teachers when I was a student.

Now, I will introduce the teaching staff distribution situation, including the architecture aspect, and the degrees we provide in city planning. First, as for architecture, it is mainly master's degree; as for city planning, we have projects for both undergraduate and postgraduate. As for geographical surveying, we have projects for postgraduate of 10 terms, as well as projects of real estate and land management, and all our postgraduate projects have corresponding requirements according to the laws and regulations of Bulgaria. All our teachers are professors and postgraduates. Meanwhile, we also have structure engineering, civil engineering, project management, etc. For me, I work for the college of structure engineering. There is college of water conservatory project, which includes water conservatory project, water supply, sewage treatment, water management, and other majors. We also have a college of traffic engineering, which offers master degree, including traffic engineering, with ten terms of teaching. After the completion of 8 terms of teaching, or precisely, after the completion of the 4th terms, as for the students of railway engineering and road engineering, they will have a special study of practical engineering. Meanwhile, we also have cooperation with Technische Universit? t Wien. These are the academic certifications of us, including the certification from Royal Institute of British Architects of various colleges of Bulgaria. This project of us takes science and practice as orientation, thus we will also organize special disciplinary conference. We organized a series of scientific research activity in the 75th-anniversary ceremony in our university. Moreover, the former president of Bulgaria also visited our university to guide our work once.

Besides, we also took part in many education projects of Europe. This is our campus, which is the campus in the middle of Sofia. This part of our campus is just for teaching activity and this is one of the teaching buildings.

For sure, we also face some problems, like the new challenges we meet. However, from now on, we should face the challenges jointly under the new initiative of the Belt and Road Architectural University International Consortium.

As for the highway on the picture, I will not explain in details. Here, we list our vision and our objective. We hope further expand our recruitment scope and carry out further operation through this conference. Thank you all for listening.

"一带一路"倡议背景下建筑类大学的国际交流合作与创新人才培养的探索与实践

吉林建筑大学校长　戴　昕

（一）"一带一路"倡议对建筑类高校发展的重要意义

2016年，教育部印发《推进共建"一带一路"教育行动》（以下简称《教育行动》）。《教育行动》提出教育交流为"一带一路"沿线各国民心相通架设桥梁，人才培养为沿线各国政策沟通、设施联通、贸易畅通、资金融通提供支撑。倡议沿线各国和地区携手行动起来，增进理解、扩大开放、加强合作、互学互鉴，谋求共同利益、直面共同命运、勇担共同责任，聚力构建"一带一路"教育共同体，全面支撑共建"一带一路"。

"一带一路"倡议有助于开辟建筑类大学国际合作与交流新方向，"一带一路"倡议贯穿欧亚非大陆，涉及中亚、西亚、南亚、东南亚、中东、中东欧、部分欧洲发达国家和非洲国家，这些沿线区域可以成为建筑类大学重点开展合作的区域，沿线国家的大学可以成为合作对象的新选择。

"一带一路"倡议有助于丰富建筑类大学国际合作与交流新内涵，建筑类大学应有效转变国际合作与交流内涵，发挥比较优势，由"引进""吸收""学习"向"输出"转变，以到沿线国家开展境外办学、开展多种形式的学位项目、吸引沿线国家学生来华留学等形式，服务"一带一路"建设。

"一带一路"倡议有助于创新建筑类大学国际合作与交流新模式，为适应"一带一路"倡议所倡导的双边、多边合作机制以及区域和次区域合作理念，建筑类大学应创新国际合作与交流的新模式，改变“一对一”的合作模式，以大学联盟的形式与沿线国家的建筑类大学开展多边的“区域联盟合作”。

（二）“一带一路”倡议背景下建筑类高校发展面临的机遇与挑战

面对“一带一路”倡议背景下建设高水平建筑类大学的发展目标，建筑类大学发展还存在一些亟待解决的问题和困难。主要表现在：综合办学实力与国际高水平大学还存在着较大差距；基础研究和原始创新能力不强，重大创新平台和标志性成果数量偏少；国际影响力亟待提升。

建筑类大学必须充分、准确地把握国内外高等教育改革的发展方向，认真遵循高等教育发展的基本规律，以高度的事业心和强烈的使命感，立足新起点，谋求新发展，开创新局面，坚定不移地加快改革发展步伐，持之以恒地推进创建具有中国特色的高水平建筑类大学进程。

大学具有人才培养、科学研究、社会服务和文化传承四大功能。建筑类大学必须担当起“一带一路”建设的思想库、智囊团和动力源功能，概括起来，就是做好“四个支撑”。

一是提供人才支撑。深化人才培养机制改革，有计划地培养有科技背景、有领导能力、熟悉沿线国家政策法律、历史文化的国际化复合型建筑人才。大力发展面向“一带一路”沿线国家的留学生教育，培养更多知华友华的高端建筑人才。

二是提供科技支撑。与“一带一路”沿线国家建立一批产学研紧密结合的国际合作联合实验室和技术转移中心，共同研究解决沿线各国共同面临的气候变化、环境保护、基础设施建设、绿色建筑等领域的关键科学技术问题。

三是提供智力支撑。围绕“一带一路”倡议，深化国际智库学术对话，加强国际建筑科技创新研究、国际建筑文化比较研究、国际建筑教育改革实践研究，促进沿线国家政策协调、战略对接和民心相通，为“一带一路”规划设计、机制创新、方向路径和精准推进做好建筑领域咨询服务。

四是提供文化传播支撑。让“一带一路”沿线国家进一步了解中国建筑文化，让中国进一步了解“一带一路”沿线国家悠久的历史文化。

（三）“一带一路”倡议背景下吉林建筑大学国际交流合作与创新人才培养模式的探索

1. 学校国际交流合作概况

我校积极推进与各国和地区大学及机构的友好合作。与美国、俄罗斯、英国、韩国、日本、新加坡、中国台湾等20多个国家和地区的60多个教育机构建立联系。在教师互访、学生互派、合作研究等方面开展国际合作与交流。积极选派优秀骨干教师赴境外留学培训。通过国家外专局、国家留学基金委等项目选派了近百名骨干教师和管理人员赴境外进行为期3～12个月的留学培训。

我校引进国外优质教育资源，实施中外合作办学项目。与美国、俄罗斯等4所大学的中外合作本科教育项目被教育部批准。我校积极推动境外办学工作，与美国、英国等3所大学达成联合培养硕士研究生意向。我校积极共建国际科研合作平台，与英国巴斯斯巴大学创意计算中心共同搭建两校国际科研合作平台，开展深层次实质性科研合作。

引进外国专家学者来校工作讲学。近5年来，我校引进英国、韩国的知名大学专家2人作为吉林省“长白山学者”特聘专家来校工作，共接待30多个国家与地区的1000余人次来宾来访，邀请境外知名专家学者来校做学术讲座及授课近100场次。引进海外高层次人才来校工作，引进获得国外著名大学博士学位的高层次人才20余人。

学生出国（境）交流学习工作快速健康发展。近5年来，我校为学生提供多种赴境外学习交流的机会，共计派出800余名学生出国（境）交流学习。积极承办国际学术会议，共主办（承办）亚洲城市环境学会国际学术会议等10次。

我校积极开展“一带一路”沿线国家来华留学工作。通过申请及设立“一带一路奖学金”“吉林省政府奖学金”“吉林建筑大学来华留学生校长奖学金”，并不断完善留学生教材建设、全英文品牌课程建设，保证留学生招生和人才培养质量。近年来，已有100余名来自18个沿线国家的留学生来校学习。

2. 国际工程管理人才教育

我校以国际工程管理专门人才培养为试点，实现学校专业教育、注册执业资格教育与国际工程教育接轨，探索培养具有国际工程视野、通晓国际工程规则、能够参与国际工程事务和国际工程竞争的国际工程管理人才的培养模式和途径。

1990 年，学校选派优秀中青年骨干教师参加原国家建委和国家教委联合主办的国际工程管理研究生班；1992 年，学校创设了国际工程管理专门化人才培养项目；2000 年，学校实现从 Fidic（国际工程合同）单一教育向国际工程管理全方位创新教育的转变；2010 年，学校在国际工程管理专门化人才培养项目建设基础上，获批教育部国际工程管理双学士学位试点单位；2014 年，学校进一步增设 BIM 和 FM 两个国际工程管理双学士学位试点方向；2017 年，学校成立全国首家 BIM 学院。

我校采用项目负责人制的管理模式，保障项目的顺利开展与实施；在全校大土建背景专业范围内选拔专业基础好、英语能力强的优秀人才，保证生源质量；将项目全部收费划拨给试点团队自由支配，学校每年固定划拨 40 万元作为项目建设专项经费。

通过为双语教师支付高倍酬金的方式，有效吸引高水平师资，保障项目高水平教学质量；建立了校内实体与校外虚拟相结合的教学团队；形成了“以我为主、引智为辅”的国际工程管理教学团队；开发了基于互联网的虚拟教学平台，改革了传统的教学方法；开展国际工程管理人才的订单式培养。

近年来，围绕“一带一路”倡议，我校积极探索国际工程管理人才教育教学模式；基于“一带一路”沿线国家土建类人才需求，构建国际工程管理人才的培养模式、培养方案、课程体系、质量监控与评价体系。培养了 600 余名拥有土木建筑类各专业背景、掌握国际工程项目管理专业知识和技能，具备创新能力、奉献精神、协同开拓的复合型国际工程管理人才，毕业生中绝大多数就职于涉外建筑企业，深受用人单位的好评。

学校连续三年在由中国建设教育协会主办、国家住房和城乡建设部高等教育工程管理专业指导委员会协办的“全国高等院校学生斯维尔杯 BIM 大赛”中，共获得 2 个全能二等奖、3 个全能三等奖、4 个专项二等奖、12 个专项三等奖、1 个参赛院校组织一等奖的优异成绩；在连续四届全国“龙图杯”BIM 大赛高校组中，学校取得了 2 个一等奖、4 个二等奖、1 个三等奖的好成绩。

学校国际工程管理人才教育的探索与实践推动了全校工程类专业的教育教学综合改革，得到了清华大学、哈尔滨工业大学、重庆大学、商务部培训学院非洲访问团等国内外知名专家、学者、官员的高度认可和评价，国内相关院校也多次来学校调研、考察与交流。

（四）“一带一路”倡议背景下吉林建筑大学发展趋势与展望

1. 进一步明确学校国际合作与交流发展思路

吉林省是国家确定向北开放的枢纽，在“一带一路”倡议中具有独特的地位与作用。目前，吉林省正努力将长吉图战略、东北亚区域开放合作战略融入“一带一路”建设，努力围绕打造“丝路吉林”大通道，加大铁路、公路、桥梁、口岸、基础设施等规划建设力度，提高内外联通水平。

学校作为吉林省唯一的土建类高等学校，以落实“一带一路”倡议作为整个学校国际合作与交流工作的主线，指导学校各学科、各院部的国际交流合作，不断拓展与“一带一路”沿线国家和地区交流合作规模、层次与水平。依托大学联盟，主动融入国家战略和吉林省战略，主动作为，为“一带一路”沿线、国家和地方经济社会发展做出应有的贡献。

2. 进一步实现优势互补形成合力

“一带一路”沿线国家建筑类大学分布广、数量大、底蕴足，发展特色鲜明。学校依托联盟，遵守联盟章程，贯彻执行联盟决议，与联盟各成员间广泛交流，实现资源共享，促进共同发展，实现优势互补，形成发展合力，不断提高联盟间的协同育人、科技攻关等的能力和水平，为“一带一路”倡议实施提供建筑领域内强有力的人才支撑、科技支撑、智库支撑、文化支撑。

3. 进一步发挥学校复合型国际工程创新人才培养功能

我校以培训、长短期留学、国际会议、国际论坛等为途径，以改革人才培养模式、人才培养方案、课程设置、教学评价体系，加强师资队伍、实习实训基地、网络课程建设等为抓手，不断提高具有国际视野、知华友华、掌握建筑专业技能与管理知识的复合型国际工程人才的培养质量与水平。

4. 进一步发挥学校科技创新功能

围绕“一带一路”沿线国家和地区技术需求，充分发挥学校学科和科研整体优势，加强与“一带一路”沿线国家和地区政府、大学及“走出去”的中国企业密切合作，建设校政、校校、校企联合科技研发基地，积极开展国际科技合作，不断产生“适时、适地、适用”的创新科技成果。

5. 进一步发挥学校社会服务功能

积极推进与“一带一路”沿线国家和地区共建科技成果转化平台，加大科技成果转化力度，促进科研成果惠及“一带一路”沿线国家与人民，进而全面提升学校服务国家、服务社会、服务人民的能力与水平。

6. 进一步发挥学校文化传承功能

围绕“民心相通”，秉持“文化自信”，增强与沿线国家和地区青年的人文交往与文化交流，增进与沿线国家和地区人民之间的理解和互信，发挥学校的文化传播与文化传承功能。

（五）结语

“一带一路”倡议将为建筑类大学的建设与发展提供宽广的舞台，“一带一路”建筑类大学国际联盟将为建筑类大学搭建深化国际合作与交流的平台。建筑类大学将持续创新人才培养，深化国际交流与合作，在服务国家战略需求、服务国家发展和实现“一带一路”共赢中作出新的更大的贡献！

Exploration and Practice of International Exchange and Cooperation & Cultivation of Innovative Talents in Architectural Universities under The Belt and Road Strategy

Dai Xin
President of Jilin Jianzhu University

I. Significance of the Belt and Road Strategy for the Development of Architectural Universities

The Ministry of Education issued the *Education Action for Promoting Joint Construction of the Belt and Road* (J. W. [2016] No. 46), which points out that educational exchange serve as a bridge among people from all countries along the Belt and Road, while talent cultivation provides support for policy coordination, connectivity of infrastructure, unimpeded trade, financial integration, and closer people-to-people ties among such countries; it calls for all the Belt and Road countries to work together, enhance mutual understanding, expand opening up, strengthen cooperation, learn from each other, seek common interests, face common destiny, bear common responsibilities, form synergies to build the Belt and Road education community, and support the Belt and Road construction in all aspects.

(1) The Belt and Road strategy is supportive of opening up new directions for international cooperation and exchanges among architectural universities; as the strategy covers Eurasia and Africa, involving Central Asia, West Asia, South Asia, Southeast Asia, the Middle East, Central and Eastern Europe, some Euro-

pean developed countries as well as African countries, these regions along the route could become the key areas for cooperation among architectural universities, and universities in these countries could become new options for cooperation.

(2) The Belt and Road strategy facilitate to enrich the contents of international cooperation and exchanges among architectural universities; architectural universities should effectively transform the contents of international cooperation and exchanges, give full play to their comparative advantages, change from "input", "absorption" and "learning" to "output", so as to carry out overseas education in the countries along the Belt and Road, carry out various forms of academic degree programs, attract students from these countries to study in China, thus serving the implementation of the Belt and Road strategy.

(3) The Belt and Road strategy helps to innovate new models of international cooperation and exchanges among architectural universities. To adapt to the bilateral and multilateral cooperation mechanisms, as well as the regional and sub-regional cooperation concepts advocated by the Belt and Road Initiative, architectural universities should innovate models of international cooperation and exchanges, change the "one-to-one" cooperation model, and carry out multilateral "regional consortium cooperation" with architectural universities in the Belt and Road countries by forming alliances or consortia.

II. Opportunities and Challenges Facing the Development of Architectural Universities Under the Belt and Road Strategy

Facing the development goal of building high-level architectural universities under the Belt and Road strategy, there are still some problems and difficulties to be solved in the development of architectural universities, mainly reflected as: big gap of comprehensive educational level with high-level international universities; weak strength in basic research and original innovation, few major innovation platforms and landmark achievements, and weak international influence in urgent need for improvement.

Architectural universities shall fully and accurately follow the development direction of higher education reform both in China and abroad, strictly observe the basic rules of higher education development, cultivate a strong sense of dedication and mission, set off from the new starting point, seek for new development, create a new situation, unnervingly step up reform and development, and persistently promote the establishment of high-level architectural universities with Chinese characteristics.

University possesses four functions, namely, talent cultivation, scientific research, social service and cultural inheritance; architectural universities shall serve as a talent pool, think tank and power source for the Belt and Road construction; to put it simply, we shall provide "four supports".

Firstly, providing talent support. Efforts shall be made to deepen the reform of the personnel cultivation mechanism, and systematically cultivate international inter-disciplinary architectural talents with scientific and technological background, leadership skills, familiar with policies, laws, histories and cultures of the Belt and Road countries. We shall vigorously develop education for international students from such countries and cultivate more high-end architectural talents who know well about and are friendly with China.

Secondly, providing technological support. Efforts shall be made to set up a batch of international cooperation joint laboratories and technology transfer centers that are closely integrated with industry, research and universities from the Belt and Road countries, so as to jointly research and solve key scientific and technological issues in areas such as climate change, environmental protection, infrastructure construction, green building faced by all the countries along the route.

Thirdly, providing intellectual support. We shall focus on the Belt and Road Initiative, deepen academic exchanges of international think tanks, strengthen research on international architectural science and technology innovation, intensify comparative study of international architectural culture, facilitate research on international architectural education reform practices, promote policy coordination, strategic connection, and people-to-people ties of countries along the Belt and Road, and provide consultation services in architectural planning and design, mechanism innovation, direction path and precise progress of the Belt and Road Initiative.

Fourthly, providing support for cultural communication. We shall enable countries along the Belt and Road to further understand the Chinese architectural culture, and in turn, help China to learn more about the time-honored histories and cultures of these such countries.

Ⅲ. Exploration on the Model of International Exchange and Cooperation & Innovative Talent Cultivation by Jilin Jianzhu University (JJU) Under the Belt and Road Strategy

(1) Overview of JJU's International Exchanges and Cooperation

JJU actively promotes friendly cooperation with overseas universities and institutions. We maintain relationships with more than 60 educational organizations in more than 20 countries and regions such as the United States, Russia, the United Kingdom, South Korea, Japan, Singapore and Taiwan; carryout international cooperation and exchanges in areas such as teacher visits, student dispatch and cooperative research. We also actively select outstanding backbone teachers to go abroad for training, as nearly 100 backbone teachers and management staff have been sent abroad for 3-12 months of training sessions under the projects launched by the State Administration of Foreign Experts Affairs, China Scholarship Council, etc.

JJU introduces high-quality foreign educational resources and implement international cooperative education projects. The undergraduate education project cooperated with four foreign universities including the United States and Russia have been approved by the Ministry of Education of China. We actively promote overseas education, and have reached cooperative intentions with three universities in the United States and the United Kingdom for joint cultivation of graduates. We also actively build the international scientific research cooperation platforms, and have established the international scientific research cooperation platform with the Computing Center of Bath Spa University, the United Kindom, to carry out in-depth and substantive cooperation for scientific research.

JJU introduces foreign experts and scholars to work and give lectures in China.

In the past 5 years, we have invited 2 specially-appointed experts under the tile of Jilin "Mount. Changbai Scholars" from well-known universities in the UK and South Korea to work in our university, received over 1, 000 visitors from more than 30 countries and regions, and conducted almost 100 academic lectures and courses given by invited well-known overseas experts and scholars. We also introduce overseas high-level talents to work in our university, and have introduced more than 20 high-level talents with doctoral degrees from renowned foreign universities.

JJU speeds up the sound development of overseas exchange and study for students. In the past 5 years, we have offered students with a variety of opportunities to study and exchange abroad, sending a total of more than 800 students overseas. We also actively undertake international academic conferences, sponsoring (organizing) 10 international academic conferences of the Asian Urban Environmental Society.

JJU actively carries out the work for organizing foreign students from the Belt and Road countries to study in China. We apply and set up the "Belt and Road Scholarship", "Jilin Provincial Government Scholarship", "JJU President Scholarship for International Students", and constantly improve the construction of international students' teaching materials and all-English courses, so as to ensure the education quality for international students. In recent years, more than 100 international students from 18 Belt and Road countries have come to study here.

(2) Education of Talents in International Engineering Management

Through trial cultivation of talents for international engineering management, JJU seeks for integrating its academic education, registered qualification education with international engineering education, and explores models and measures for cultivating international engineering management talents with international vision, proficient in international engineering rules, and capable to participate in international engineering affairs and competitions.

In 1990, JJU selected excellent young and middle-aged backbone teachers to participate in the graduates' class of international engineering management co-sponsored by the former State Construction Commission and the National Education Commission; in 1992, JJU launched the project of specialized talent cultivation for international engineering management; in 2000, it transformed from the mono FIDIC education to an all-round innovative education in international engineering management; in 2010, based on special talent cultivation projects for international engineering management, JJU was approved by the Ministry of Education as a pilot unit for double bachelor's degrees in international engineering management; in 2014, it further launched two trial programme of double bachelor's degrees in international engineering management of BIM and FM; in 2017, it established China's first BIM college.

JJU adopts the management mode of leader-responsible system to ensure the smooth development and implementation of its projects; we select excellent talents with good academic foundation and proficient English language skills with civil engineering background to ensure the quality of students; we also allocate all the fees to the pilot teams for free disposal, as JJU allocates RMB 400,000 *yuan* per year as a special fund for project development.

By paying higher salaries for bilingual teachers, JJU effectively attracts high-level teachers to guarantee the high-level teaching quality; establishes the teaching teams that combine on-campus with off-campus entities; forms the international engineering management teaching team featuring "relying on our own talents while supported by introduced talents"; develops a virtual teaching platform based on the internet, reforms traditional teaching methods; carries out order-based talent cultivation in international engineering management.

In recent years, we have proactively explored the education and teaching model of international engineering management by centering on the Belt and Road Initiative, and built the model for international engineering management talent cultivation, cultivation planning, curriculum system, quality monitoring and evaluation system based on the demand for civil engineering talents from countries along the Belt and Road. We have cultivated more than 600 inter-disciplinary international engineering management talents who have various professional backgrounds in civil engineering and construction, academic knowledge and skills in international engineering project management, innovative ability, as well as the dedicated and collaborative development spirit. Our graduates are widely engaged in foreign-related construction enterprises and are well recognized by their employers.

JJU has achieved remarkable performances for three consecutive years in the "National Sverd Cup BIM Competition for Colleges and Universities" sponsored by the China Construction Education Association and co-sponsored by the Higher Education Project Management Steering Committee of the Ministry of Housing and Urban-Rural Development, wining a total of 2 all-round second prizes and 3 all-round third prizes, 4 special second prizes, 12 special third prizes, and 1 first prize for participating colleges and universities. In addition, we have also won 2 first prizes, 4 second prizes and 1 third prize in national "Longtu Cup" BIM competition for colleges and universities.

The exploration and practice of JJU's international engineering management talent education has promoted the all-round reform of education and teaching for engineering majors throughout the university, and has been highly recognized and evaluated by well-known experts, scholars, and officials from both Chinese and abroad, such as from Tsinghua University, Harbin Institute of Technology, Chongqing University, and African delegations at the training school of Ministry of Commerce; meanwhile, related domestic universities have also frequently visited our university for investigation, inspection and exchange.

IV. Development Trend and Prospect of JJU Under the Belt and Road Strategy

(1) Further Clarifying Ideologies of International Cooperation and Exchange Development

Identified by China as the hub opening to the north, Jilin Province boasts a unique position and role in the Belt and Road strategy. At present, Jilin Province is endeavoring to integrate its Changchun-Jilin-Tumen River strategy and the opening-up cooperation strategy of Northeast Asia into the Belt and Road strategy, and underline the planning of railways, highways, bridges, ports and infrastructure in its efforts of building great passages of the "Silk Road Jilin", so as to improve internal and external connectivity.

As the exclusive university of civil engineering in Jilin Province, JJU follows the main line of implementing the Belt and Road Initiative for international cooperation and exchange, guides international exchange and cooperation of various disciplines and departments, and continuously expands the scale, scope and level of exchanges and cooperation with countries along the Belt and Road. Relying on the consortium, JJU also actively integrates into the national strategies and Jilin Provincial strategies, takes initiative to make proper contributions to the Belt and Road Initiative, as well as to national and local economic and social development.

(2) Further Forming Synergy with Complementary Advantages

Along the Belt and Road, there are extensive and multiple architectural universities with distinctive characteristics. JJU relies on the BRAUIC, abides by its articles of association, implements its resolutions, communicates extensively with its member universities, facilitates resource sharing, promotes common development, enable the complementation of advantages, forms synergy for development, and constantly improves its capability and level in conducting cooperative education and making technology breakthroughs with BRAUIC members, and provides strong support in talent, technology support, think tank and culture for the implementation of the Belt and Road strategy.

(3) Further Developing the Function of Cultivating Inter-disciplinary Creative Talents in International Engineering

JJU takes measures such as training, long-and short-term overseas study, international conferences, international forums; reforms the talent cultivation model, talent cultivation program, curriculum arrangement and teaching evaluation system; strengthen the building of the teaching faculties, internship & practice bases, as well as online courses, thus continuously improving the quality and level of cultivating inter-disciplinary international engineering talents who know well about and are friendly with China, and possess international visions, architectural professional skills and management knowledge.

(4) Further Developing the Function of Technological Innovation

Focusing on the technological needs of countries along the Belt and Road, JJU gives full play to its overall academic and scientific research strengths, intensifies cooperation with governments, universities and Chinese enterprises along the Belt and Road, builds R&D technology bases for cooperation with governments, universities and enterprises, actively carries out international scientific and technological cooperation, and continuously yields "timely, appropriate and applicable" innovative scientific and technological achievements.

(5) Further Developing the Function of Social Service

JJU actively pushes forward the establishment of the platform for transforming scientific and technological achievements with countries along the Belt and Road, enhances such transformations, and promotes scientific research achievements to benefit countries and people along the Belt and Road, so as to comprehensively improve our ability to serve the country, the society and the people.

(6) Further Developing the Function of Cultural Inheritance

Focusing on "people-to-people bond" and adhering to "cultural confidence", JJU enhances people-to-people and cultural exchanges among youths from countries along the Belt and Road, enhances mutual understanding and trust, and exerts the university's functions in cultural communication and inheritance.

V. Closing Remarks

The Belt and Road strategy provides a grand stage for the construction and development of architectural universities, and the BRAUIC would serve as a platform for architectural universities to deepen international cooperation and exchanges. Architectural universities should continue to cultivate innovative talents, deepen international exchanges and cooperation, and make new and greater contributions in serving the country's strategic needs and development, and achieving win-win results for the Belt and Road Initiative!

“一带一路”发展，未来都市建设与创新人才培养

韩国大田大学校长　李钟瑞

大家好，我是韩国大田大学的校长李钟瑞。

首先，我对北京建筑大学举办的“一带一路”建筑类大学联盟开幕式表示衷心祝贺，并对张爱林校长和王建中书记，以及北京建筑大学各位老师的邀请表示感谢，同时也感谢国内及海外大学各位来宾的莅临。

大田大学坐落于韩国的大田广域市，大田广域市距离韩国的首都首尔有一个小时的路程，位于韩国国土的正中央，是韩国的交通枢纽、科学中心和行政中心，同时也是今年成立的新政府指定的第四次产业革命特别城市，大田可以说是一个未来型都市。

大田大学是一所建校于1980年的四年制本科院校，7个学院中有本科在校生10000多人，硕士研究生和博士研究生1000多人在读。大田大学被韩国政府选为“产学合作先导大学培养项目（LINC+）”“本科教育先导大学培养项目（ACE）”和“地方大学特性化项目（CK-1）”三个领域的政府财政支援学校，被韩国教育部和法务部评选为“教育国际化能力大学”。

本次在北京建筑大学设立的“一带一路建筑类大学联盟”是积极响应中国住建部、外交部和商务部公布的《丝绸之路经济带21世纪海上丝绸之路的规划实践》以及国家“一带一路”倡议的项目。北京建筑大学与世界各国大学的共同参与，意味着相互间的未来发展更值得期待。

我们通过历史可以了解到，从两千多年以前到产业革命的实现，丝绸之路对东西方之间的经济、文化交流合作起了重大作用。虽然产业革命的实现，

以及国际化、数字化的世界让我们在时间上、空间上都变得更加亲近，但另一方面，理念与宗教的差异也引起了不少矛盾，国家与国家之间的经济差距也有所拉大。我认为，今天我们通过一起建设符合新时代要求的丝绸之路，来追求东西方之间的合作、共存、和平与共赢，这是时代的需求。如果是为了寻求这种具有崇高意义的地区合作的话，我们大田大学将会不惜一切所具备的人力资源、知识信息来积极参与这个项目。

特别是针对北京建筑大学促进的“一带一路”事业，大田大学将会利用在建筑、土木、环境方面的优秀业绩作为后盾，在两校间的学生、教授的交流，乃至产学合作的关系的发展上不断做出努力。另外，大田大学也会积极参与作为第四次产业革命都市的“大田广域市”在智能城市与 IT 领域的建设。文化背景不同的人才聚集在一起，对教育与研究的探讨，本身就是时代要求的创新性人才培养的最佳方案。

2015 年，在我校副校长的带领下，包含研究生院院长在内的各个负责人已经访问了贵校，2016 年北京建筑大学党委副书记和各位老师也访问了我校。通过相互访问，两校已经签署了合作协议，并进行了研究生的交流活动。通过这次访问，相信两校之间的合作会更加具体、更加坚固，同时也会扩大与海外会员大学的合作。

再次对北京建筑大学举办的“一带一路建筑类大学联盟”表示祝贺，并对张爱林校长和王建中书记以及各位老师表示诚恳的感谢，并预祝今天参加本次会议的各大学在相互合作方面繁荣发展。谢谢！

The Development of the Belt and Road, the Construction of Future Cities and the Cultivation of Innovative Talents

Lee Jong Seo, President of Daejeon University, South Korea

Hello everyone, I am the president of Daejeon University Lee Jong Seo,

First, I sincerely express my congratulations on "The Opening Ceremony of the Belt and Road Architectural University International Consortium" sponsored by Beijing University of Civil Engineering and Architecture and thanks president Zhang Ailin, secretary Wang Jianzhong, and all teachers of Beijing University of Civil Engineering and Architecture for inviting me, as well as thanks the guests from domestic government organs and overseas universities for visiting.

Daejeon University locates in Daejeon City of South Korea, with only one hour journey from Seoul, the capital city of South Korea. Locating in the center of South Korea, it is a transportation hub, scientific center and administration center, as well as the special city of the 4th industry revolution designed by the newly established government. Therefore, Daejeon city can be called as a future city.

Daejeon University is four-year undergraduate university, established in 1980. There are totally more than 10,000 undergraduate in all the 7 colleges, more than 1000 postgraduates and doctor students. Daejeon University has been selected as the financial supporting university in the three fields of "Leading University Cultivation Project (LINC+) of Industry and Study Cooperation", "Leading University Cultivation Project (ACE) of Undergraduate Education", and "Special Project

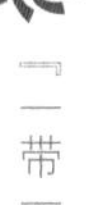

(CK-1) of Local Universities" by South Korea government, rated as "University of International Education Ability" by ministry of education and ministry of justice.

"The Belt and Road Architectural University International Consortium" set up by Beijing University of Civil Engineering and Architecture is to respond the *Planning Practice of the Silk Road Economic Belt and the 21st Century Maritime Silk Road issued by the* MOHURD, MOFA, MOFCOM of China as well as the project of "the Belt and Road" development strategy of the state. The joint participation of Beijing University of Civil Engineering and Architecture and various universities in the world means that future development among each other deserves better expectation.

From history, we can understand that the Silk Road has had a great impact on the economic and cultural exchange between the east and the west from 2000 years ago to the realization of industry revolution. Though the realization of industry revolution, internationalization and the digital world make us closer in time and space, the difference in ideas and religions also causes a lot of contradictions, and the economic gap between countries is widening. I think it is the needs of the era for us to build the Silk Road to meet the requirements of the new era to pursue cooperation, co-existence and peace and mutual benefit. If it is to look for regional cooperation with this noble significance, our Daojeon University will take part in this project positively with all the human resources and knowledge information.

Especially for the Belt and Road course promoted by Beijing University of Civil Engineering and Architecture, Daejeon University will utilize its excellent performance in architecture, civil engineering, environment as the backup force, and make continuous efforts for the exchange of students and professors between the two schools as well as the development of industry and study cooperation. Moreover, Daejeon University will also participate in the smart city and IT field construction of "Daejeon City" which is the 4th industry revolution city. As for the discussion on education and research itself, the gathering of talents of different culture backgrounds is the best solution for innovative talent cultivation required by the era.

In 2015, led by the vice president of our university, each responsible person including the president of the college of postgraduates visited your school. In 2016, the vice Party secretary and many teachers of Beijing University of Civil Engineering and Architecture also visited our university. The two universities have signed cooperation agreement through mutual visits and carried out exchange activity of postgraduates. Through this visit, I believe the cooperation between two schools will be more specific and firmer, and the cooperation with other overseas member universities will also be expanded.

I express my congratulations on "the Belt and Road Architectural University International Consortium" sponsored by Beijing University of Civil Engineering and Architecture again, as well as my sincere appreciation for president Zhang Ailin, secretary Wang Jianzhong and all teachers, and sincerely wish the prosperous development of mutual cooperation among all the universities participating in this conference today. Thank you!

借力“一带一路”倡议，全面推进高校教育国际化

沈阳建筑大学校长　石铁矛　陈宗胜

尊敬的张爱林校长，加吉克·加斯蒂安校长，同志们：

大家下午好！

今天我们相聚北京建筑大学，为了共同的愿景——搭建建筑教育信息共享、学术资源共享的交流合作平台，探索跨国培养与跨境流动的人才培养机制，促进联盟高校之间校企的双向流动，成立了“一带一路”建筑类大学国际联盟。“一带一路”是促进共同发展、实现共同繁荣的合作共赢之路，下面我结合“一带一路”倡议，与大家分享沈阳建筑大学的教育国际化发展之路。

在“一带一路”国际合作高峰论坛上，国家主席习近平明确提出，要推动“一带一路”沿线国家教育合作，扩大互派留学生规模，提升合作办学水平。我校紧紧围绕“一带一路”倡议，明确对外交流与合作、留学生教育工作任务，增强国际化工作意识，开拓创新对外交流与合作，全面推进学校教育国际化建设工作。

（一）积极开拓国际合作渠道，整合吸收国外优质教育资源，不断开展实质性的国际联合办学

“一带一路”倡议提倡和平合作、开放包容、互学互鉴、互利共赢，学校积极拓展国际交流与合作，以实质性项目合作建设国际教育文化发展共同体。目前我校与40多个国家和地区的80余所大学建立了校际交流关系并进行实质性的合作，搭建了与美国迈阿密大学、英国谢菲尔德大学、澳大利亚

悉尼科技大学、意大利米兰理工大学等世界知名学府交流的平台，构建了俄罗斯莫斯科国立建筑大学、圣彼得堡国立建筑大学、顿河国立技术大学等行业特色明显的高校合作群体，打造了俄罗斯、波兰、罗马尼亚、捷克、克罗地亚等“一带一路”沿线国家和地区的中东欧高校特色合作体。基于校际合作关系，我校以骨干教师培训交流为切入点，结合中外优质教育资源，推动合作办学工作建设，实现以项目合作带动校际关系深层次发展。

学校坚持“引进来”与“走出去”并重、学术交流与技术研究共荣的教师交流机制。紧紧依托国家和省关于高层次人才公派出国留学计划，每年选派骨干教师赴国外培训交流30余人次，在技术前沿、跨学科领域与国外专家通力研究合作。以教师交流为契机，注重学校重点项目或重要科研课题外国专家引智，学校成功获得国家外专局高端外国专家项目和国家、省级重点引进国外专家智力项目20余项，有效地活跃了学校科学研究的氛围，不断提高学校的核心竞争力。其中，依托中英种植基地申报的英国谢菲尔德大学景观学院院长詹姆斯教授作为外专“千人计划”人选项目已通过初评。

在学术互信、技术共通的基础上，学校不断引进国外优质教育资源，促进中外合作办学工作的全面开展。一方面，进一步巩固和规范中美、中德、中俄和中英硕士及本科合作办学与交流项目，提升中外合作办学的层次，加强省委组织部高层次人才培训基地的建设，努力打造示范性中外合作办学经典。自2005年以来，中美信息管理硕士学位合作办学项目已经为辽宁省委组织部培养各级各类干部10期329名，为社会各行业培养学员13期529名，总计858名；有一大批学员走上了县（处）级领导岗位，20余名同志被提拔到副厅级以上领导岗位。另一方面，学校积极探索与俄罗斯、罗马尼亚等“一带一路”沿线国家高校合作举办建筑学、土木工程和建筑与环境专业硕士及本科层次的中外合作办学项目，整合土建类专业优势，借力国家“一带一路”倡议，进一步扩大学校中外合作办学的规模，不断提升学校的国际化办学水平。

（二）推进学校课程体系国际化建设，扩大学生互派交流规模，实现高层次人才国际化培养

“一带一路”倡议的全面推进，人才培养是关键。高校作为国家高等教育主体，高层次人才培养更是学校永恒的使命。我校从拓展学生的国际视野、增强学生参与国际竞争的能力、提高学科专业人才培养水平和质量这三点出发，进一步明确国际化人才培养要求，科学确立国际化的课程目标，进一步加强课程设计国际化、课程内容国际化，强化课程教学过程国际化，并积极参加课程国际认证，从课程、师资、管理、留学、学分互认、学位互授等方面全方位建设国际化办学体系。

基于国际化办学体系建设，在学生互派交流工作中，学校从短期交流着手，向学位交流深入，实现双学位深化。我校每年选派优秀硕士生和本科生赴美国、德国、俄罗斯、罗马尼亚、捷克等11个国家和地区的16所国（境）外高校参加的交流项目有20个，学位项目有百余个，形成了美国伊利诺伊大学芝加哥分校、澳大利亚悉尼科技大学、意大利米兰理工大学等世界一流学府引领，校际学费互免，食宿费资助、奖学金等财力支持，课程学分互认、规章制度等系统保障的良好工作局面。同时借助“一带一路”倡议，重点引进俄罗斯、罗马尼亚等国家合作院校学生来校进行专业学习，建立专业学习与研究实践相互促进的学研结合模式。其中，罗马尼亚特来西瓦尼亚大学交流项目得到欧盟认同，获批伊拉斯谟世界项目计划资助，德国维斯马大学双学位硕士互换交流项目更是获得德意志学术交流中心（DAAD）奖学金资助，业已开展15年，两校互换交流学生达到200余名。

（三）建立国际科研平台，促进学术研讨交流，全面推动国际科技合作

结合当前“一带一路”沿线国家和地区的产业开发利用和综合治理并举的阶段性特色，利用建筑类高校科技与行业优势，我校注重联合创新，攻坚技术瓶颈，构建人才培养培训、产学研合作以及行业前瞻发展三位一体的国际科研平台，通过实质性项目共建，实现技术交流长效发展。

目前，我校与芬兰坦佩雷应用科技大学合作的北欧木结构节能环保别墅

示范中心已经完成并投入使用。同时，在辽宁省政府、沈阳市政府和德国驻华大使馆的支持下，德国外交部“德中同行”在我校启动的后续项目“中德节能示范中心”也已竣工投入使用。中德节能示范中心获得国家三星级绿色建筑认证和德国能源署“节能创新奖”。这两个国际合作示范中心的建成，为开展“绿色、低碳、节能、生态”建筑节能技术的发展研究提供了良好的国际科研合作平台。

学校积极建立国家级建筑节能环保国际示范工程项目研究、应用和培训基地，消化吸收国际先进的建筑节能技术，推进寒地低碳建造技术的发展。积极探索与“一带一路”沿线国家和地区高水平大学和科研院所开展国际合作研究，注重基础性、前沿性和成果转化并推动学科建设发展。努力搭建国际学术交流合作载体，抓住时机、创造条件举办与创办高层次国际学术会议和高水平国际论坛，不断提高学校的国际竞争力。

（四）弘扬中国语言及文化，加强海外办学建设，快速推动孔子学院发展

“国之交在于民相亲，民相亲在于心相通”。近年来，孔子学院建设快速发展，已经成为加强中国人民与世界各国人民友谊合作的桥梁。我校全面考虑国际合作院校特点以及孔子学院全球布局，选取了“一带一路”沿线国家罗马尼亚为第一所孔子学院建设点。2012 年 3 月 26 日，我校和罗马尼亚特来西瓦尼亚大学共同建立了第一所孔子学院。目前孔子学院在校学生数量达到 1000 余名，并在开门办学、外地设点办学、大学设立汉语本科专业和特色文化活动等方面不断突破，设立了教学点 14 个，在罗开设了中国语言和文学本科专业，获批中国国家汉语国际推广领导小组办公室（下文简称国家汉办）“孔子学院奖学金项目”。先后举办“一带一路，人文交流”等系列罗马尼亚孔子学院夏令营活动，实现了学校海外办学工作的顺利开展。此外，我校启动了另一个“一带一路”沿线国家波兰的孔子课堂建设布局，已经完成了与波兰琴希托霍瓦工业大学共同建立独立孔子课堂向国家汉办的申请工作，并力争在 2017 年底正式获得国家汉办批准。

（五）建立国外留学生长效发展机制，突出留学生教育特色，全面发展来华留学生工作

学校在充分发挥学校学科专业特色与优势的基础上，紧跟国家政策导向，进一步扩大国际留学生教育规模。2012 年学校被教育部批准成为“接受中国政府奖学金来华留学生资格”的高校。2014 年学校获批辽宁省政府外国留学生博士研究生奖学金资格。基于“一带一路”国家高校合作基础，我校申请并获批教育部中东欧学分生专项奖学金项目，接收罗马尼亚学生来校学习。借助国家“丝绸之路”留学推进计划，我校成为首批“丝绸之路”中国政府奖学金委托培养留学生高校，吸引了大批俄罗斯学生来校学习。同时，由于“一带一路”地缘优势，学校留学生教育在巴基斯坦树立了品牌，巴基斯坦学生成为学校留学生重要组成部分。学校在土木、机械、信息和工商管理等专业领域开设了英文授课的留学生招生专业，留学生教育覆盖本、硕、博三个层次，留学生规模快速增长，留学生增量连续三年位列辽宁省高校第一位，今年各类在校留学生已经超过 650 人。

学校将充分发挥学校优势与特色，多渠道多形式地开展长、短期外国留学生教育工作，不断扩大外国留学生规模，大力发展学历教育。重点抓好师资和留学生管理两支队伍建设，建立健全来华留学生教育质量评估体系，不断扩大学校国际影响力。

高校的教育国际化发展之路历来是机遇与挑战并存，“一带一路”倡议是教育国际化跨越式发展的历史性契机，需要加强领导与监督，结合工作实际，统筹规划，加强顶层设计，分解具体任务，强化推进措施，确保教育国际化各项工作任务落到实处。

谢谢大家！

Availing the Belt and Road Strategy to Promote Internationalized Higher Education in all Aspects

Report at the Founding Ceremony of the BRAUIC

Shi Tiemao[1] Chen Zongsheng*

Shenyang Jianzhu University

Distinguished President Zhang Ailin, President Gagik Galstyan, and Comrades,

Good afternoon!

Today we gather at Beijing University of Civil Engineering and Architecture (BUCEA) for a common purpose, that is, to build the exchange and cooperation platform for sharing architectural education information and academic resources, explore the mechanism for cross-border talent cultivation and mobility, and promote the two-way flow between enterprises and consortium member universities, and found the Belt and Road Architectural University International Consortium (BRAUIC). The Belt and Road brings about win-win cooperation for common development and common prosperity. Now, I would like to share with you the experiences of Shenyang Jianzhu University (SJU) in facilitating education internationalization by availing the Belt and Road strategy.

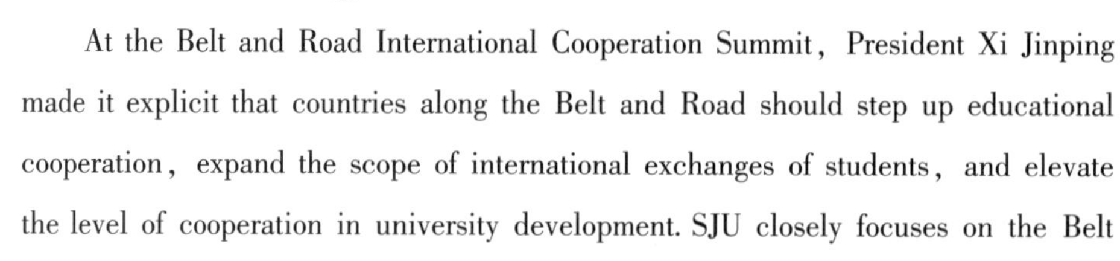

At the Belt and Road International Cooperation Summit, President Xi Jinping made it explicit that countries along the Belt and Road should step up educational cooperation, expand the scope of international exchanges of students, and elevate the level of cooperation in university development. SJU closely focuses on the Belt and Road Strategy, identifies the tasks of overseas exchange and cooperation, as well as the education of international students, enhances the awareness of internationalization, innovates overseas exchanges and cooperation, and comprehensively promotes internationalization of higher education.

1. Actively exploring channels for international cooperation, integrating and absorbing high-quality foreign educational resources, and continuing substantive international joint education

Under the call of the Belt and Road Initiative for peaceful cooperation, openness & accommodation, mutual observation, mutual benefit for win-win results, SJU actively engages in expanding international exchanges and cooperation, and building the international education and cultural development community through substantive projects. Currently, it has built inter-university exchange relations and carried out substantive cooperation with more than 80 universities in over 40 countries and regions. We have set up a platform for exchanges with world renowned universities such as University of Miami, the United States; The University of Sheffield, the United Kingdom; University of Technology Sydney, Australia, and Polytechnic University of Milan, Italy. We have also established the cooperative university clusters with obvious characteristics such as Moscow State University of Architecture, St. Petersburg State University of Architecture, and Don River State Technical University, and has created a featured cooperative entity with universities in the Central and East European countries along the Belt and Road such as Russia, Poland, Romania, Czech Republic and Croatia. Based on the inter-university partnership, SJU takes backbone teacher cultivation and exchanges as the starting point, combines Chinese and foreign high-quality educational resources, promotes joint education, and realizes in-depth development of inter-university relations through project cooperation.

SJU sticks to develop the teacher exchanging mechanism which lays emphasis on both introducing in and going out, as well as on academic exchange and technical research. Relying closely on the national and provincial plans for funding overseas study of high-level talents, over 30 backbone teachers are sent to overseas for training and exchanges each year to cooperate with foreign experts in researching cutting edge technologies and interdisciplinary study. Availing the opportunities of teacher exchanges while focusing on the introduction of foreign experts in key university projects or important scientific research subjects, we have successfully applied for more than 20 high-level foreign expert projects from the State Administration of Foreign Experts Affairs as well as the national and provincial introduction of key foreign experts intelligence projects, thus have effectively enlivened the atmosphere of scientific research and constantly improved the university's core competitiveness. Among them, via application by the Sino-British planting base, Professor James, Dean of the Landscape College of The University of Sheffield, the United Kingdom, has passed the preliminary evaluation as a candidate for the "Thousand Talents Program" for foreign experts.

Replying on mutual academic trust and technology sharing, SJU has continuously introduced in high-quality foreign educational resources to promote cooperative education between China and foreign countries in all aspects. Efforts have been made to further consolidate and standardize the Sino-US, Sino-German, Sino-Russian and Sino-British undergraduate and graduate cooperative education and exchange programs, elevate the level of cooperative education, strengthen the building of high-level talent cultivation bases under the Provincial CPC Committee Organization Department, and strive to create exemplary cooperative education cases. Among them, since 2005, a total of 858 personnel have been trained, including 329 cadres at all levels of all types for Liaoning Provincial CPC Committee Organization Department trained under the the Sino-US Information Management Master's Degree Cooperative Education Program, and 529 personnel trained in 13 sessions for various sectors of the society. A large number of trainees have taken up leadership positions at the county (division) level, and over 20 comrades have been promoted to leadership positions at or above the deputy department level. Meanwhile, efforts have also been made to actively explore cooperation with Russia, Romania and other countries along the Belt and Road in organizing undergraduate and graduate joint education in architecture, civil engineering and architecture & environment, integrating advantages of the civil engineering disciplines, avail the country's Belt and Road strategy to further expand cooperative education among universities in China and abroad, and continuously elevate our international level of education.

2. Promoting the construction of internationalized curriculum system, expanding student exchange, and realizing internationalized cultivation of high-level talents

Talent cultivation is crucial to the all-round implementation of the Belt and Road strategy. As the main body of national higher education, university shoulder the eternal mission of high-end talents cultivation. Starting from expanding students' international horizons, enhancing students' ability in international competition, and improving the level and quality of talent cultivation, we have further clarified the requirements for international talent cultivation, scientifically established international curriculum objectives, and further strengthened internationalized curriculum design and teaching process, actively participated in international certification of courses, and built an all-round international education system in courses, teachers, management, overseas study, mutual recognition of credits, and mutual award of degrees.

Based on the construction of the internationalized university system, in the process student exchanges, SJU starts with short-term exchanges, deepens the student exchange for achieving double degrees. Every year, outstanding master and undergraduate students are sent to engage in the exchange or degree programs in 16 universities (overseas) in 11 countries and regions including the United States, Germany, Russia, Romania, Czech Republic, etc., thus shaped a sound pattern featuring guidance by world-class universities such as University of Illinois at Chicago, University of Technology Sydney, Australia, and Polytechnic University of Milan, Italy, financial support of mutual tuition exemption, subsidies for accommodation and scholarships, as well as mutual academic credit recognition and institutional guarantees. Furthermore, availing the global strategic advantage of the Belt and Road Initiative, we have also introduced students in especial from cooperative universities in Russia, Romania and other countries, and set up the academia-research model of combining academic learning and research practices. Among them, the exchange program of University of Transylvania, Romania has been recognized by the European Union and funded by the Erasmus World Project Program; moreover, the double-degree graduate exchange program at the University of Wismar, Germany, has also been awarded the scholarship from the German Academic Exchange Center (DAAD), enabling exchanges between the two universities for over 200 students in the past 15 years.

3. Establishing the international scientific research platform, promoting academic discussions and exchanges, and facilitating international scientific and technological cooperation in all aspects.

Based on the current phased features of industrial development & utilization and comprehensive governance of countries along the Belt and Road, SJU takes the technological and industrial advantages of architectural universities, focuses on joint innovation, technical bottlenecks, builds the international scientific research platform featuring integration of talent cultivation, industry-university-research cooperation and prospective industrial development, and realizes long-term development of technological exchanges through substantive projects.

Currently, the Nordic wood structure energy-saving and green villa demonstration center co-built with Tampere University of Applied Sciences, Finland, has been completed and put into operation. Meanwhile, under the support of the Liaoning Provincial Government, Shenyang Municipal Government and the German Embassy in China, the follow-up project "Sino-German Energy Conservation Demonstration Center" under the "Germany and China" programme initiated in our university by the German Ministry of Foreign Affairs has been completed and put into operation. The Sino-German Energy Conservation Demonstration Center has won the national three-star green building certification and Innovative Energy-saving Award of the German Energy Agency. These two international cooperation demonstration cen-

ters have served as good international scientific research cooperation platform for research on "green, low-carbon, energy-saving and ecological" architectural technologies.

SJU actively establishes national-level bases for the research, application and training of architectural energy conservation and environmental protection international demonstration engineering projects, absorbs internationally advanced architectural energy-saving technologies, and promotes the development of low-carbon construction technology in cold regions. We actively explore international cooperative research with high-level universities and research institutes in the Belt and Road countries, focus on the transformation of basic and cutting-edge achievements, facilitate the development of discipline construction, strive to build a carrier of international academic exchanges and cooperation, seize opportunities, create favorable conditions to host and launch high-level international academic conferences and forums, and continuously improve the university's international competitiveness.

4. Promoting Chinese language and culture, strengthening overseas education and stepping up the development of Confucius Institutes

State-to-State relationship lies in people-to-people friendship, which furthermore depends on mutual communication. In recent years, the rapid development of the Confucius Institute has served as a bridge to strengthen the friendship and cooperation between people from China and the world. Upon comprehensive considerations of the characteristics of universities under international cooperation as well as the global strategic layout of the Confucius Institute, SJU has selected Romania from the Belt and Road countries as the first site for setting up the Confucius Institute, as on March 26, 2012, SJU and University of Trevania, Romania, jointly founded the first Confucius Institute, which has, up till now, recruited more than 1, 000 students. We has also made continuous breakthroughs in carrying out education in other places, provided the Chinese Language major for undergraduates, launched characteristic cultural activities, set up 14 teaching points and opened Chinese Language and Chinese Culture major for undergraduates in Romania, successfully applied for the Confucius Institute Scholarship Program from The Office of Chinese Language Council International, organized a series of Confucius Institute Summer Camp activities themed on "Belt and Road Humanities Exchange" in Romania, and has made smooth development of overseas university education. In addition, we have also made arrangement for another Confucius Classroom in Poland, and has filed the application to The Office of Chinese Language Council International for setting up an independent Confucius Classroom with Poland's Censhitohova University of Technology, and are striving to obtain its official approval by the end of this year.

5. Establishing a long-term development mechanism for overseas students, highlighting the educational characteristics for international students and enabling them to study in China.

While making full utilization of the characteristics and advantages of the university's disciplines, SJU closely follows national policies to further expands education for international students. In 2012, SJU was approved by the Ministry of Education for its qualification of sponsoring international students with Chinese government scholarship; in 2014, it was approved by the Liaoning Provincial Government for its qualification of sponsoring international students for doctoral scholarship. Based on the cooperation between universities under the Belt and Road Initiative, we have successfully applied from the Ministry of Education for special scholarship for Central and Eastern European credit students to recruit Romanian students to study in SJU. Under the national "Silk Road" international students promotion plan, SJU became one of the first batch of universities to educate foreign students with the "Silk Road" Chinese government scholarships, which has attracted a large number of Russian students. Meanwhile, thanks to the geographical advantages of the Belt and Road Initiative, our education for international students becomes a good brand in Pakistan, as Pakistani students make up an important part of the university's international students. We have opened majors instructed in English language for foreign students of civil engineering, machinery, information and business manage-

ment, and provided education for undergraduates, graduates and doctors. The scale of international students in our university has grown rapidly, as the increment of our international students has ranked first among universities in Liaoning Province for three consecutive years, and this year alone has witnessed over 650 international students of various kinds.

We will give full play to the university's advantages and characteristics, carry out long and short-term education for international students in multiple ways and forms, continuously enlarge the population of foreign students, and vigorously develop academic education. We will also focus on the development of teachers and management of foreign students, establish and improve the educational quality evaluation system for international students in China, and continuously expand the our international influence.

The work to internationalize university education has always been full of opportunities and challenges. Facing the historic opportunities brought about by the Belt and Road strategy for leap-forward development of internationalized education, we should strengthen leadership and supervision, start from realities, coordinate planning, improve top-level design, break down specifics tasks, intensify measures for advancement, and ensure all tasks of internationalized education to be earnestly implemented.

Thank you for your attention!

在“一带一路”建筑类大学国际联盟校长论坛上的报告

俄罗斯建筑土木科学院院士　帕维尔·阿基莫夫

各位尊贵的来宾，大家下午好！我非常高兴能参加“一带一路”建筑类国际大学联盟校长论坛。我们学校并不是一个教育院所，而是一个科学研究院所，但我们和俄罗斯境内以及国际的各大学都有着广泛的合作，今天我想跟大家非常简短地介绍一下我们开展的活动。

首先，我和大家一起回顾一下历史。我们机构相对比较年轻，大概在26年前，主要是在当时的总统支持下成立的。这并不是俄罗斯第一所建筑土木科学院，第一所是于18世纪建立、由雕塑建筑和绘画领域的三所院校合并产生的。我们科学院的院长是俄罗斯建筑工程学院的一位院士，也是我们国内非常有名的一位建筑师，在此之前，他也是莫斯科城的首席建筑师。

其次，我跟大家介绍一下，我们单位主要有三种成员：全职成员、通信会员和国外会员。我们各成员都是通过选举产生的，标准主要是基于他们在科学方面的贡献。目前我们全职成员有57位，还有不少国外的成员，这也是基于他们的科学贡献评选出来的。在这张PPT上，大家可以看到，我们学院的管理层，有不少都是非常杰出的科学家和工程师。比如其中一位，是我们当地一座标志性建筑的工程师；有一位院士建造了19世纪圣彼得堡久负盛名的世界最高建筑之一。关于我们研究院，在这里向大家介绍一下：第一，我们进行了科学的预测，特别是在建筑、土木工程和城市规划方面。第二，积极参与俄罗斯联邦各项工作，以推动国家可持续发展和建设。第三，为俄罗斯和其他区域建筑类高校提供学术支持，并且也积极地参与地区性的国际交

流与合作。关于国际交流与合作，包括建筑领域、规划和市政建设方面的不少国际知名机构。在这张 PPT 上面，可以看到我们开展的不少活动。比如我们在应用和基础研究方面的参与和协调，对教育机构提供支持，其中包括了俄罗斯的一些高校，还有海外的一些高校。除此之外，我们在建筑、市政规划方面得到了俄罗斯政府的支持，借助这些我们举办了不少的国际合作、国际会议和国际交流的活动，因为时间关系，不能向大家一一列出。

此外，我想稍微跟大家介绍一下我们的出版物和一些学术著作。我们主要科学研究的期刊叫做《建筑建设科学》学术期刊，这个期刊里面会刊出不少论文，都是我们会员单位一些研究的成果汇总，其中有不少都颇具影响力。另外一份出版物叫做《国际计算机土木市政工程》国际期刊，这个期刊是由我们的合作伙伴 AE three 作为编辑出版的，他今天出席了本次活动。这一学术期刊，主要关注的就是建筑科学方面的计算机建模和数学建模。在这张 PPT 上，主要介绍了我们机构单位，主要分为三个部分：第一，是关于市政规划部、建筑部，还有建造科学部等，每个下设部有具体的部门细分。第二，我们还有 38 个分支机构，位于我们国家的中部、南部、西北部、远东、西伯利亚、克里米亚地区，还有区域中心，也是今年刚刚成立的机构。第三，理事会和委员会都有特定的使命，比如我们有学术委员会、一般性学术委员会、建筑专业委员会、建造科学委员会，他们专注于不同领域，有的是关注市政规划、文物遗址的保护和修复，有的是具体的建造技术，有的是土壤科学，还有和环保生态相关的结合学科，我的同事明天将会就这个内容做更加具体的发言。我的另外一位同事，明天会和大家更详细地介绍一下，我们科学委员会关于科技、计算机建模还有数学建模在建筑行业当中的应用。我们最重要的合作伙伴，是国际土木工程学会，在这个学会中，150 多位会员是来自俄罗斯的机构或者个人成员，有不少还是来自原苏联加盟共和国。这张 PPT 主要向大家介绍我们科学院获得的奖项，其中一些奖项，获得者不仅包括了教职员工，还有表现杰出的本科生和研究生。我们对于国家项目的参与，也是非常广泛的。在这张 PPT 上面大家可以看到，我们也参与了俄罗斯国家重点工程，在 2013 年到 2020 年俄罗斯联邦科学技术发展规划当中，我们也做出了自己长期的贡献。

接下来的几张 PPT 主要向大家介绍我们对基础研究做出的贡献，主要包括三个方面的内容：建筑、建设和城市规划。第一是关于建筑和城市规划当中理论和历史问题的解决和研究，第二是关于建筑和城市规划交叉学科的研究，第三是关于一些市政规划基础性的理论研究。除此之外，还要对未来研究的趋势进行预测，并且要对建筑科学的最新进展提供理论支持，下面列出了不少分支领域，比方说建材方面，还有工程力学方面等。

我们还有一个研究方向，就是要保证对环境和城市宜居的考量，换言之，不仅要考虑技术本身，更多的是要考虑在当下环境中的建筑对宜居性和环境的兼容性。

最后，再向大家介绍一下我们开展的一些科学活动。我们研究院已经在国内外组织了 100 多个不同的研讨会和学术活动。对于我来说，比较重要的就是建筑计算机建模的一系列研讨会。我们在俄罗斯境内已经召开了六届研讨会，第七届研讨会将在明年举行，而且也是在俄罗斯知名的一家土木工程学院进行，我也诚邀在座各位同事，届时能够积极参加会议。

今天，我借这个 PPT 向大家介绍一下我们在计算机结构建模领域开展的工作，大家可以看到，我们在过去已经举办了不少次学术会议，下一期将是在 2018 年 2 月份的时候举行。我们还和出版社合作开展了一些工作，我们出版的内容包括了国内最杰出的行业当中的教材，还有一些辅导类材料，相信在未来会有更多优秀的教材出版。我们也非常期待着能够和我们“一带一路”建筑类大学国际联盟的各个成员单位进一步加强合作。非常感谢大家的聆听。

Report on the Belt and Road Architectural University International Consortium and Presidents' Forum

Pavel Akimov, Academician of Russian Academy of Architectural and Civil Sciences

Distinguished guests!

Good afternoon!

I am very happy to attend the President Forum of "the Belt and Road Architectural University International Consortium" . We do not belong to an educational institution, but we belong to a scientific research institution. Despite this, we have extensive cooperation with various universities in Russia and international universities. In my speech, I would like to introduce to you very briefly about our activities today.

Firstly, let's review the history of our organization. Our institution is relatively young, which was probably established 26 years ago. It was mainly established with the support of the President. Despite this, it is not the first Russian civil engineering school. The first building civil engineering institute was established in the 18th century. It was mainly produced by the merger of three institutions in the field of sculpture architecture and painting. You can also see the dean of our academy, who is also an academician of the Russian Academy of Architecture and Engineering. In addition, he is also a very famous architect in our country. Before that, he was also the chief architect of Moscow City.

Secondly, let me introduce the three types of members that our unit mainly includes: full-time members, communication members and foreign members. Our various members are elected, mainly based on their contribution to science. Currently we have 57 full-time members. In addition, we have many foreign members, which are based on their scientific contributions. You can also see on this PowerPoint that many of our college management are very outstanding scientists and engineers. For example, one of them is an engineer in one of our local landmarks. In addition, one of our institutions has built one of the world's most prestigious buildings in St. Petersburg in the 19th century. The main objectives introduced here are: first, our predictions for science, especially for construction, civil engineering and urban planning; second, we actively participate in the work of the Russian Federation to promote the country, especially the sustainable development and construction; third, it provides academic support for Russia and regional architecture universities, and actively participates in regional international exchanges and cooperation. Regarding international exchanges and cooperation, it includes internationally renowned institutions such as construction, planning and municipal construction. From this PowerPoint, we can see that we have carried out a lot of activities: first, our participation and coordination in application and basic research; second, we provide support to educational institutions, including some universities in Russia. It also includes some colleges and universities overseas. In addition, we have received support from the Russian government in terms of construction and municipal planning. With these, we have organized many activities on international cooperation, international conferences and international exchanges. Because of the time limited, I can't list them one by one.

In addition, I would like to briefly introduce you to our college publications and some academic works. Our main scientific research journals are called Academic Journal of Construction Science. In this journal, there are a lot of published papers, which are a collection of some research results of our member units, many of which are quite influential. Another publication is called International Journal of Computer Civil and Municipal Engineering, which was published by our partner AE Three. He also attended the activity today. In addition, the main focus of this academic journal

is computer modeling and mathematical modeling in architectural science. On this PowerPoint, we mainly introduce the units of our organization. There are three main units: the municipal planning department, the construction department, and the construction science department. Each unit is subdivided into specific departments. Second, we have 38 branches in the central, southern, northwest, Far East, Siberia, and Crimea regions of our country. Volga and the regional center are the institutions that have just been established this year. Third, the council and the committee have specific missions. For example, we have academic committees, general academic committees, architectural professional committees, and construction science committees. On this PowerPoint we can see our scientific committees, which focus on different areas, such as municipal planning, protection and restoration of cultural relics, concrete construction techniques, some soil science, and environmental protection. There are also integrated disciplines related to environmental protection and ecology. My colleague will make a more specific statement on this content tomorrow. In addition, I would like to briefly introduce the application of technology, computational modeling and mathematical modeling in the construction industry. In this regard, my other colleague will also give you a more detailed introduction in the speech tomorrow. In addition to this, there are committees on geological earthquake science. Our most important partner is the International Civil Engineering Society. In this society, more than 150 members are from Russian institutions, and many of them are from the former Soviet Union. From this PowerPoint, you can see the awards won by our Academy of Sciences, some of which are awarded to faculty and staff, as well as outstanding undergraduate and graduate students. In addition, our participation in national projects is also very extensive. As you can see from this PowerPoint, we are involved in the Russian National Priority Project. In the Russian Federation's science and technology development plan from 2013 to 2020, we have made our long-term contribution.

The next few PowerPoint mainly introduces our contribution to basic research, which mainly includes three different contents: about architecture, about construction, and about urban planning. The first is about the solution and research of theoretical and historical issues in architecture and urban planning; the second is about the interdisciplinary study of architectural urban planning; and the third is the theoretical research on the basics of some municipal planning.

In addition, we need to predict the trends of future research and provide theoretical support for the latest developments in building science. The following PowerPoint also lists a number of branch areas, such as building materials, engineering mechanics and so on.

The eighth research direction is to take into account environmental and urban livability considerations. In other words, we not only need to considerate the technology itself, but also the current environment for the compatibility of livability and environmental compatibility.

Finally, let me introduce you to some of the scientific activities that we have carried out. Our institute has organized more than 100 different conference seminars and academic events. The location is not only at home but also abroad. For me, what's more important is a series of workshops on computational modeling in architecture. We have held six seminars in Russia. The seventh seminar will be held next year at a well-known civil engineering institute in Russia. I also invite all colleagues here to actively participate in this conference.

I use this PowerPoint to briefly introduce our work on computer structural modeling. As you can see from this, we have already held many such academic conferences in the past. The next issue will be held in February next year. This is some of the work we and the publishing house have done. The content we publish basically includes the most outstanding materials in our industry, as well as some materials for counseling. We believe that we will have more outstanding materials published in the future, and we are also looking forward to further strengthening cooperation with the member organizations of the Belt and Road Architectural University International Consortium. Thank you very much for listening.

“一带一路”绘就建筑类高校发展新蓝图

安徽建筑大学校长　方潜生

2013 年 9 月 7 日，中国国家主席习近平访问哈萨克斯坦，提出要共同建设“丝绸之路经济带”，并于同年 10 月在出访东南亚国家期间，提出共建“21 世纪海上丝绸之路”的重大倡议。“一带一路”倡议是习近平同志深刻思考人类前途命运以及中国和世界发展大势，为促进全球共同繁荣、打造人类命运共同体所提出的宏伟构想和中国方案，是习近平新时代中国特色社会主义思想的有机组成部分，开辟了我国参与和引领全球开放合作的新境界。

（一）“一带一路”倡议背景下，建筑类高校的机遇与挑战

1. “一带一路”建设为建筑类高校带来的机遇

《推动共建海上丝绸之路经济带和 21 世纪海上丝绸之路的愿景与行动》明确提出，基础设施互联互通是“一带一路”建设的优先领域。

目前各地“一带一路”拟建、在建基础设施规模已经超过 1 万亿元，跨国投资规模约 524 亿美元，且投资项目多为铁路、公路、机场、水利建设；2015 年“一带一路”基建投资项目总规模已经达到 1. 04 万亿元，其中铁路投资近 5000 亿元，公路投资 1235 亿元，机场建设投资 1167 亿元，港口水利投资金额超过 1700 亿元。

2013 年，“一带一路”建设带来的海外收入占我国八大建筑公司总营收的 16%，极大提升了我国大型建筑公司的海外收入占比，沿线国家中，已经有近 60 个国家明确表示支持和积极参与建设。

2015年上半年，中国企业在“一带一路”沿线60个国家承包工程项目达1401个，直接投资总额达70.5亿美元。2016年度全国建筑业总产值20余万亿元，比上年同期增长7.1%；从业人员超过5000万，是国民经济重要的支柱产业。据世界银行预测，2016年到2030年，全球基础设施投入需要将近50万亿美元，相当于每一年3.3万亿美元。据测算，未来十年，每年亚洲基础设施的投资规模约8000亿美元，建筑行业市场空间巨大。

这为中国建筑企业特别是基础设施建设企业带来了跨越式发展的良好机遇。同时作为建筑类高校，在“一带一路”建设发展上，有着天然的学科和专业优势，能够有更多的机会服务“一带一路”建设。

2. “一带一路”为建筑类高校带来的挑战

“一带一路”建设的瓶颈不是资金，也非技术，而是人才。基础设施建设人才尤为缺乏，我国建筑业现有人才，对项目的操作经验还仅限于国内的市场环境，甚至大部分只有在本省本地区才有生存的可能，而“一带一路”现在锁定的更多是基础设施建设的项目，这些项目投资大、周期长，很多项目前期的科研、论证、设计还不到位。

（二）建筑类高校服务“一带一路”建设的优势——以安徽建筑大学为例

1. 发挥区位优势，共建“一带一路”美好未来

2014年9月12日，《国务院关于依托黄金水道推动长江经济带发展的指导意见》定位合肥市为长三角世界级城市群副中心，世界第六大城市群副中心；2015年3月28日，《推动共建丝绸之路经济带和21世纪海上丝绸之路的愿景与行动》发布，合肥市成为“一带一路”重要节点城市，两大国家战略双重覆盖，使合肥市逐渐成为开放型经济发展的“桥头堡”。

近几年，合肥市先后获批国家第一个科技创新试点城市、综合性国家科学中心，安徽省在中国创新大格局中占据了重要地位，成为代表国家参与全球科技竞争与合作的重要力量，成为国际创新网络的重要组成部分，为中国科技长远发展和创新型国家建设提供有力支撑。

“一带一路”建设中基础设施互联互通的重大历史机遇，以及对建筑类人才和科技的迫切需求，为地处合肥的安徽建筑大学提供了重大机遇。

2. 坚持走打好“建”字牌、做好“徽”文章的特色发展之路

安徽建筑大学是安徽省唯一一所以土建类学科专业为特色的多科性大学，始建于 1958 年，是安徽省与住房城乡建设部共建高校、教育部本科教学工作水平评估优秀院校、省级博士学位授予权立项建设单位、国家“卓越工程师教育培养计划”实施高校、国家节约型公共机构示范单位、徽派建筑研究重要基地。学校设有 12 个学院，60 个本科专业，涵盖工、管、理、艺、文、法、经七大学科门类。在校全日制本科生 18000 余人，研究生 1000 余人。

我校现有 7 个一级学科、32 个二级学科硕士学位授权点，2 个专业学位授权类别，7 个专业学位授权点，8 个省级重点学科，1 个国家级工程实验室、4 个省级重点实验室、11 个国家级及省部级工程（技术）研究中心；并在节能环保、城镇化与徽派建筑、地下工程、公共安全、先进建筑材料五大重点支持的研究领域取得显著成果；依托新建科研平台，我校将重点支持新型建筑工业化、轨道交通、海绵城市、智慧城市、城市管理等新兴交叉学科领域，主动适应和服务区域经济发展，进一步推动“一带一路”建设背景下建筑类高校的发展。

3. 安徽建筑大学“一带一路”合作交流概况

我校先后与德国德累斯顿工业大学、美国伊利诺理工学院、韩国韩瑞大学、美国中密歇根大学、英国斯旺西大学、美国布里奇波特大学、白俄罗斯布列斯特国立技术大学签订校际合作协议，加强人才交流与合作，为“一带一路”的建设输送了大量国际化人才。

2015 年，学校作为全国七所牵头高校之一，联合全国其他六所建筑类大学共同组建了“一带一路”国家 A 级综合调研团；同年，“一带一路”综合调研团获“丝路新世界·青春中国梦”全国大学生“圆梦中国”优秀团队。近几年，“一带一路”综合调研团奔赴上海市、江苏省南京市、浙江省宁波市、陕西省西安市、福建省福州市、四川省成都市、重庆市等地开展调研，形成较为专业的调研报告并报告团中央。

安徽建筑大学虽然在人才培养、科学研究、学科建设、合作交流等方面取得了丰硕的成果，但是还存在着不足与需要改进的地方。因此，成立“一带一路”建筑类大学国际联盟，对推动“一带一路”沿线建筑类大学交流合

作，创新建筑类人才培养机制，探索跨文化建筑类学科专业建设，推进科技成果转化，服务“一带一路”沿线及欧亚地区的发展建设具有重大意义。

我们要借此契机，做到以下几方面：

培养高素质技能型人才。坚持科教融合，打造国际化师资队伍，创建国际教育交流学院，积极推进双向留学，培养适应“一带一路”基础项目建设的人才队伍。

提升基础研究。充分发挥高校学科综合等优势，把基础研究提升到服务“一带一路”建设的高度。

加强双边协作。要与“一带一路”沿线国家与地区畅通科研合作渠道，通过共建合作平台，共同应对“一带一路”沿线国家面临的问题和挑战，推进智库建设。利用多学科交叉研究，开展“一带一路”政策研究和咨询，特别是要对“一带一路”建设未来 5 年、15 年、50 年的发展做出科学研判，为“一带一路”建设提供具有前瞻性的指导意见。

“一带一路”倡议所倡导的双边、多边合作机制、区域和次区域合作理念，为大学国际交流与合作新模式（区域联盟合作）提供了深化发展的契机，而联盟的成立更能进一步发挥建筑类高校特色优势、推进资源共享、加强协同创新、促进人才培养，提升国际科研合作水平，提升国际交流合作的能力与水平，让我们携手共进，推动互利共赢，在“一带一路”建设背景下，绘就建筑类高校新蓝图！

The Belt and Road Helps to Bring about a New Blueprint on the Development of Architectural Colleges and Universities

Fang Qiansheng, President of Anhui Jianzhu University

On September 7, 2013, Chinese President Xi Jinping proposed for the first time that "Silk Road Economic Belt" should be built together in Kazakhstan, and during his visits to Central and Southeast Asian countries in September and October of the same year, he made major initiatives to build "Silk Road Economic Belt" and "21st Century Maritime Silk Road". The construction of the "Belt and Road" is a grand conception and a Chinese plan put forward by Comrade Xi Jinping so as to promote the common prosperity of the whole world and build a community of common destiny for all mankind after he thought deeply about the future and destiny of mankind as well as the development trend of China and the world. It is an organic part of Xi Jinping's thoughts on socialism with Chinese characteristics for a new era, and it opens up a new realm for China to participate in and lead the global open cooperation.

Ⅰ. Opportunities and Challenges of Architectural Colleges and Universities under the Background of the "Belt and Road"

1. Opportunities of Architectural Colleges and Universities Brought by the "Belt and Road"

Vision and Actions on Jointly Building Silk Road Economic Belt and 21st Centu-

ry Maritime Silk Road has made it clear that infrastructure connectivity is a priority for the construction of the "Belt and Road" .

At present, the infrastructure of the "Belt and Road" to be built and under construction has exceeded 1 trillion yuan, and cross-border investment has reached $ 52.4 billion mainly in railway, highway, airport and water conservancy construction. The total scale of infrastructure investment projects for the "Belt and Road" reached 1.04 trillion yuan in 2015, wherein 500 billion yuan in railways, 123.5 billion yuan in highways, 116.7 billion yuan in airport construction, and over 170 billion yuan in port water conservancy.

In 2013, the overseas income brought by the "Belt and Road" strategy accounted for 16% of the total revenue of China's 8 major construction companies, greatly increasing the proportion of the overseas income of China's large-scale construction companies. Nearly 60 countries along the Belt and Road have clearly expressed their willing to support and actively participate in the construction.

In the first half of 2015, Chinese enterprises contracted 1,401 projects in 60 countries along the "Belt and Road", and the total direct investment reached $ 7.05 billion.

The total output value of the construction industry in the country was over 20 trillion yuan in 2016, an increase of 7.1% over the same period of last year. The number of employees exceeded 50 million. The construction industry was an important pillar industry of the national economy.

The World Bank predicts that from 2016 to 2030, nearly $ 50 trillion will be required by global infrastructure investment, equivalent to $ 3.3 trillion per year. Estimates show that Asia's infrastructure investment in the next decade will be about $ 800 billion each year, which will bring about huge market space for the construction industry.

This brings a great opportunity for China's construction enterprises, especially infrastructure construction enterprises, to achieve leapfrog development. At the same time, as architectural colleges and universities, we have natural discipline and professional advantages and can have more opportunities to serve the "Belt and Road" during the construction and development of the "Belt and Road" .

2. Challenges of Architectural Colleges and Universities Brought about by the "Belt and Road"

The bottleneck of the "Belt and Road" strategy is not capital, nor technology, but talents, especially talents in infrastructure construction. China's existing construction talents only have project operation experience under the domestic market environment, even most of whom can achieve survival only in the province or region, while the "Belt and Road" now has more infrastructure construction projects, which have huge investment and long cycle. Many of the projects have improper research, demonstration and even design in the early stage.

Ⅱ. Advantages of Architectural Colleges and Universities in Serving the Construction of the "Belt and Road" - Taking Anhui Jianzhu University as an Example

1. Make full use of location advantages and jointly build a better future for the "Belt and Road"

On September 25, 2014, *Guiding Opinions of the State Council on Promoting the Development of the Yangtze River Economic Belt by Relying on the Golden Waterway* located Hefei as the vice-center of the world-class urban agglomeration in the Yangtze River Delta, the world's sixth-largest urban agglomeration. On March 28, 2015, *Vision and Actions on Jointly Building Silk Road Economic Belt and 21st Century Maritime Silk Road* located Hefei as an important node city of the "Belt and Road" . Covered by the two national strategies, Hefei gradually becomes the "bridgehead" of the development of the open economy.

In recent years, Hefei has been approved as China's first pilot city for scientific and technological innovation and comprehensive national science center, and Anhui has taken an important position in China's innovation pattern, become an important force to participate in global science and technology competition and cooperation on behalf of the country and an important part of the international innovation network, and provided strong support for the long-term development of China's science and technology and the construction of an innovative country.

Facing the major historical opportunities of infrastructure connectivity in the construction of the "Belt and Road" as well as the urgent demand for architectural talents and science and technology, Anhui Construction University located in Hefei has been provided a great opportunity.

2. Insist on the development road characterized by "Architecture" and "Anhui" marks

Founded in 1958, Anhui Jianzhu University is the only multi-disciplinary university featured in civil engineering in Anhui Province. It is a joint university with the Ministry of Housing and Urban and Rural Construction. It is a university co-founded by Anhui Province and Ministry of Housing and Urban-Rural Development, an outstanding university for the evaluation of undergraduate teaching work level by Ministry of Education, a provincial project construction unit entitled to grant doctorate, a university implementing national "Educational Training Program for Excellent Engineers", a demonstration unit of national conservation-oriented public institutions and an important research base of Anhui style architecture. Anhui Jianzhu University is composed of 12 colleges, offering 60 undergraduate programs that cover the fields of engineering, management, science, art, literature, law and economics. There are approximately 18,000 undergraduate students, of which over 1,000 are graduate students.

Our school now has seven first-level disciplines, 32 master's degree programs of the second-level disciplines, 2 professional degree program categories, 7 professional degree programs, 8 provincial key disciplines, 1 national engineering laboratory, 4 provincial key laboratories, 11 national and provincial engineering (technology) research centers. Significant results have been achieved in five key research areas, that is, energy conservation & environmental protection, urbanization &

Anhui style architecture, underground engineering, public security, advanced building materials. Relying on the newly-built scientific research platform, our school will focus on supporting new cross-disciplinary fields like new building industrialization, rail transit, sponge city, smart city, urban management and so on, actively adapt to and serve the regional economic development, and further promote the development of architectural colleges and universities under the background of the "Belt and Road".

3. Overview of the "Belt and Road" Cooperation and Exchanges of Anhui Jianzhu University

Our school has signed cooperation agreements with the Dresden University of Technology of Germany, Illinois Institute of Technology of the U. S., Hanseo University of Korea, Central Michigan University of the U. S., Swansea University of the U. K., University of Bridgeport of the U. S., and Brest State Technical University of Belarus, to strengthen talent exchanges and cooperation and supply a large number of international talents for the construction of the "Belt and Road".

In 2015, the school, as one of the seven leading universities in the country, established the "Belt and Road" National A-level Comprehensive Investigation Team jointly with six other construction universities in the country. In the same year, the "Belt and Road" Comprehensive Investigation Team won the honor of an excellent team for national college students in "Realizing China Dream" of "Silk Road New World · Youth China Dream". In recent years, the "Belt and Road" Comprehensive Investigation Team conduct investigation in Shanghai City, Nanjing City of Jiangsu Province, Ningbo City of Zhejiang Province, Xi'an City of Shaanxi Province, Fuzhou City of Fujian Province, Chengdu City of Sichuan Province, Chongqing City and other places, formed professional research reports and sent them to the Youth League Central Committee.

Although Anhui Jianzhu University has made fruitful achievements in talent training, scientific research, discipline construction, cooperation and exchanges and so on, there are still some deficiencies and points that need to be improved. Therefore, the establishment of the Belt and Road Architectural University International Consortium is of great significance in promoting the exchanges and cooperation of architectural universities along the "Belt and Road", innovating the mechanisms of architectural talent training, exploring the construction of cross-cultural architectural disciplines, promoting the transformation of scientific and technological achievements, and serving the development and construction of countries along the "Belt and Road" and Eurasia.

We will take this opportunity to do the following:

Cultivate high-quality skilled talents. Adhere to the integration of science and education, build an international team of teaching staff, set up a school of international education exchanges, actively promote two-way study abroad, and cultivate a team of talents who adapt to the construction of the "Belt and Road" infrastructure projects.

Enhance basic research. Give full play to the advantages of the university like subject synthesis and raise the basic research to the level of serving the "Belt and Road".

Strengthen bilateral cooperation. Work with countries and regions along the "Belt and Road" to keep channels of scientific research cooperation open and jointly address the problems and challenges faced by countries along the "Belt and Road" by jointly building cooperation platforms.

Promote the construction of think tanks. Carry out policy research and consultation on the "Belt and Road" through multidisciplinary cross-disciplinary research, in particular make scientific research and judgments on the development of the "Belt and Road" construction in the next 5 years, 15 years and 50 years, and provide forward-looking guidance for the "Belt and Road" construction.

The bilateral and multilateral cooperation mechanisms, regional and sub-regional cooperation concepts advocated by the "Belt and Road" provide the opportunity to deepen development for universities in international new exchange and cooperation model (regional alliance cooperation), and the establishment of the alliance can further give play to the advantages of architectural colleges and universities, promote resource sharing, enhance collaborative innovation, promote talent training and improve the level of international research and cooperation and enhance the ability and level of international exchanges and cooperation. Let's work together to promote mutual benefit, and draw a new blueprint on the development of architectural colleges and universities under the background of the "Belt and Road"!

“一带一路”建筑类大学国际联盟，前所未有的新机遇

法国马恩·拉瓦雷大学副校长　弗里德里克·杜马泽

尊敬的张爱林校长，尊敬的各位来宾、各位朋友：

我很高兴今天能够来到这里，遗憾的是，我们的校长没有办法来跟大家一起进行讨论，我就代表我的学校来进行发言。

我先简要介绍一下我们学校的情况，接下来再探讨一下我们在“一带一路”建筑类大学国际联盟下有哪些畅想。首先跟大家分享一下，我们校长给张爱林校长发了祝贺函，真诚地祝贺“一带一路”建筑类大学国际联盟的成立。这是一个很伟大的成就，因为我们能够把这么多的大学汇聚到一起，共同加深学术界的合作。

接下来跟大家讲讲，我们马恩·拉瓦雷大学（简称 UPEM）的情况。大家可以看到，学校位置离巴黎不是很远，乘火车 20 分钟的车程，距离机场非常近，而且我们地理位置非常好，就在巴黎城中心和城郊迪士尼乐园中间。

我们学校拥有悠久的历史，有七大学科。我们学校主要任务是教育和科研，优势是给学生提供很多培训和实践的机会，同时也跟很多学校有学徒合作的项目，有近 25% 的学生参加过学徒制实践训练。我们学校有很多从事科研的教授，还有 500 多名硕士研究生。学校提供很多学位，有一些是技术学位，有一些是工程学位，绝大多数是硕士学位，一共有 54 个本科学位、90 个硕士学位，绝大多数都是包含了学徒制实践的。我们也非常积极地开展国际合作，加入伊拉斯默的项目，这是欧洲最大型的欧洲内外教育合作项目；我们与伙伴学校有 200 多个合作协议，还会进一步扩大合作范围。在我们新

的联盟下，我们会进一步扩大合作。下一步，我们会考虑推出一些联合的学位，这方面的工作正在开展。

我们也会借助联盟来发现一些联合学位的机会。我们和中国的合作也是很有渊源的，我们与北京、上海、武汉、包头、盐城等市都有合作，与北京建筑大学也有合作，我们还参与了中法的 HVAC 项目，我们很高兴参与这些合作，现在正在筹备推出双学位。

我们认为，“一带一路”建筑类大学国际联盟是加强合作非常有效的工具，能帮助我们建立起一个合作的网络，增加多元化合作的项目，整合一系列的资源，能够更好地与不同的合作伙伴去开发新的合作项目。目前，我们希望能够进一步推出我们的未来项目。什么是未来项目呢？这其实是一个非常大的科研项目，主要是应对以下方面的挑战：首先就是智慧城市的建设。现在很多地方都在谈智慧城市，什么才是智慧城市呢？大家对这个概念有不同的解释和理解。我们有一个重要的研究内容，就是要为未来的城市做设计，所以我们需要把城市中的方方面面都综合到设计的过程中，而且要保证能够有一个大规模的实践方法，让我们把实验室中的实践推广到更大的范围，我们也希望能够与社会经济的伙伴共同实现产学研的对接。为了实现这个目标，我们在法国也推出了一系列合作，与北京建筑大学也开展了合作，而且也得到了张爱林校长的支持，帮助我们建立这样一个未来城市的学术领导平台，让我们连接不同的大学，共同建立学术联合体。在这里我看到很多的成员，包括一些工程类大学和实验室，而且我们也会选择本身规模不是很大、但又有非常出色科研背景的学校加入未来城市联盟，我们希望进一步扩大成员的范围，不局限于法国境内的大学。

我们的目的是能够提高我们围绕城市开展的教育和科研工作的能力，不仅于此，我们希望通过共同努力，加强我们的科研能力，并且能够把研究的成果转化成能够造福社会的成果，这一点是很重要的。

接下来我们来介绍一下未来合作的前景。我们需要有交流的项目，能够让我们去进一步扩大合作，希望让我们的大学教授能共同进行科研，并推出一些共同硕士的项目，再探讨一下怎么推出博士生的项目。我们希望能够增加博士生的人数，同时也希望能够在博士生管理方面有创新。我们鼓励更多跨学科发展，比如让建筑类的专业和企业有更好的对接，我们也希望能够更好地让我们科研的成果落地。

下一步我们希望能在各方面进一步推行我们的研究生项目，包括城市、交通、科学工程、环境，还有数学和数字通信，之前发言人有提到计算机建模和数学建模对工程的作用，这对城市管理也是至关重要的。

另外一个重要的努力方向就是组织市场和机构。当然了，社会与文化也非常重要。我们的合作也有不同的形式：首先就是充分利用应用科学研究所，这是我们学校下设的一个单位，但是我们愿意对其他单位开放。我们认为，不同的领域可以解决不同的问题，但有的时候，我们也需要外界一些帮助，因为在我们的生活环境中，不少问题具有共性，我们本着“他山之石，可以攻玉”的想法，互利互助。总体来说，我们一方面加强产学结合，另外一方面加强和其他研究机构的合作，比如与纳米科学、无机材料等先进材料方面的研究机构加强合作，并且积极探索合作模式。非常感谢！

The Belt and Road Architectural University International Consortium, an Unprecedented New Opportunity

Frederic Toumazet, Vice President of
Université de Marne-La-Vallée, France

Dear President Zhang Ailin, distinguished guests and friends,

I am very happy to be here today. Regrettably, our president was unable to come to the site to discuss with you today. I am speaking on behalf of my school.

Let me briefly introduce our situation, and then discuss our imagination under the Belt and Road Architectural University International Consortium. First of all, I would like to share with you that the congratulatory letter from our president was written to President Zhang Ailin and sincerely congratulated the establishment of the Belt and Road Architectural University International Consortium. This is a great achievement because we are able to bring together so many universities to deepen the cooperation of the academic community.

Next, let's talk to you about our school. Universitéde Marne-La-Vallée, is abbreviated as UPEM. As you can see, this location is not very far from Paris, just 20 minutes by train. And UPEM is very close to the airport. This location is very good, just in the middle of the center of Paris and the outskirts of Disneyland.

To further introduce, our school actually has a long history. It is mainly composed of seven university departments. Our main task is education and scientific research. Our advantage is to provide students with many opportunities for training and practice. At the same time, we also have apprenticeship projects with many schools. Therefore, almost 25% of the students have participated in the apprenticeship practice training. Moreover, we also have many professors and more than 500 graduate students engaged in research. Then let me introduce you further to our school situation. We will provide a lot of degrees. Some are technical degrees. Some are engineering degrees. And most of them are master's degrees. We have a total of 54 undergraduate degrees and 90 master's degrees. Most degrees include apprenticeship practices, and we are also very active in international cooperation. We have also joined the Erasmus Project, which is Europe's largest collaborative project for education both inside and outside of Europe. Moreover, we have reached more than 200 cooperation agreements with partner schools, and we will further expand the scope of our cooperation. Under the new consortium, we will further expand the scope of our cooperation. Then, in the next step, we will consider launching some joint degrees, and we are currently working on this.

Of course, we will also use the consortium to explore new opportunities, such as new joint degrees. For our cooperation with China, it is also very relevant. We have cooperation with Baotou, Beijing, Shanghai, Wuhan and Yancheng, and we also have cooperation with Beijing University of Civil Engineering and Architecture. Also, we also participated in a special training program, the Sino-French HVAC Project. We are very happy to be involved in this process and are now preparing to launch a double degree.

In my opinion, including our presidents, we believe that our Belt and Road Architectural University International Consortium is a very effective tool to strengthen cooperation and help us to establish a cooperative network and diversified cooperation projects. At the same time, we also have a series of resources to better develop new cooperation projects with different partners. From the perspective of our school, we also hope to further launch our Future Project. What is a Future Project? This Future Project is a very large research project, mainly dealing with three challenges: the first is the construction of a smart city. First, we need to define what the smart city really means. Because whenever and wherever, we are talking about smart cities now, but we must be clear about what a smart city is. Then there is a smart city with predictions. Everyone has different interpretations and understandings of this concept. At the same time, we still have an important connotation, which is to design the city for the future. Therefore, we need to integrate all aspects of the city into the design process, and we must ensure that we have a large-scale practical approach. Let us promote the practice in the laboratory to a wider scope. We also hope to be able to work with social and economic partners to achieve the connection between industry, academia and research. To achieve this goal, we have launched a series of cooperation in France. We also cooperated with President Zhang Ailin and received her support. This helped us to establish a platform for academic leadership in the future city, and allow us to connect different universities and jointly establish such an academic consortium. Here you can see that all the members are listed, including some engineering universities and laboratories. Moreover, we will also choose a school with a research background that is not very large in scale but is very good itself to join in our future urban consortium, and it is not limited to universities in France. We aim to further expand the scope of our members.

We aim to improve the education and research work in our cities. But we also hope to work together to strengthen our research capabilities and to translate the results of the university into results that benefit society.

Next, I will tell you about the future of our future cooperation. We need to have a communication project that will allow us to further expand our cooperation. We aim to enable our university professors to work together on research and to launch some joint master's programs. One of the main points is to explore how to launch a Ph. D. student project. Now let's introduce the projects we have now. We also hope to increase the number of doctoral students. At the same time, we also hope to innovate at the management level of doctor and doctoral colleges. And we also want to encourage more interdisciplinary development. For example, let our construction profession and the business community to achieve better docking. And we also hope that we can better transform the results of our research into good policy.

Next we hope to further implement our graduate programs in all aspects, including cities, transportation, scientific engineering, and environment, as well as mathematics and information communication. Previous speakers have mentioned the role of computer modeling and mathematical modeling in engineering, and this is also critical for urban management.

Another important direction is to organize the market and institutions. Of course, society and cultures are also a very important point. Then the tools we work with include several different forms: The first is to make full use of the Applied Science Institute, which is also a unit under our school. We are also willing to open it to other units because we believe that different areas can solve different problems. But sometimes, we also need some help from the outside world. Because I think that many problems have commonalities in the environment we live in. We can use the stone of other mountains to attack the jade, and help others with the help of others to achieve mutual benefit and mutual assistance. In general, we strengthen the combination of industry and education on the one hand, and strengthen cooperation with other research institutions on the other hand, including cooperation in advanced materials such as nanoscience and inorganic materials. And we actively explore cooperation models. To further strengthen cooperation, I think these two topics are very important to us. This is what everyone can see on the PowerPoint. Thank you very much!

在“一带一路”建筑类大学国际联盟校长论坛上的报告

山东建筑大学副校长　范存礼

尊敬的爱林校长，各位同仁，各位嘉宾，女士们、先生们：

下午好！

我非常荣幸受校长靳奉祥教授委托，代表山东建筑大学参加北京建筑大学发起并承办的“一带一路”建筑类大学国际联盟成立大会。听了众多国内外建筑类知名高校校长的精彩发言，很受启发，下面我汇报一下我们学校的一些基本情况。

山东建筑大学坐落在山东省济南市，山东省号称齐鲁大地、“孔孟之乡”，圣人孔子，是在山东的曲阜；孟子是亚圣，是在山东的邹城。山东还有五岳之首的泰山，黄河也穿过山东省了。济南市号称泉城，天下七十二名泉就在济南市。著名的宋代词人李清照也是济南章丘人。我们学校山东建筑大学创立于1956年，经过60多年的发展，已经形成以土木建筑为特色，以工科为主，工、理、管、文、法、农、艺等多学科交叉渗透的一所多学科大学，我们现在有57个本科专业，有60个二级学科硕士点。现在在校生是26000人，毕业生每年的就业率一直在山东省的前三位。

我们学校一直非常重视国际化办学，应该说国际化是当今世界大学发展的重要特征，存在于不同类型、不同层次、不同规模、不同地域、不同宗旨的各类大学中，“一带一路”倡议为新世纪大学高水平发展提供了新的条件和可能性。建立“一带一路”建筑类大学国际联盟，从办学体系、人才培养、课程项目、学分认证、科学研究等方面着手，与“一带一路”沿线国家大学开展广泛和全面的合作，必将能更好地促进我国建筑类大学的国际化发展。目前，围绕“一带一路”建设，我们学校在推进国际化办学方面，重点做了六个方面的工作。一是推进“一带一路”沿线国家的高层互访交流，2016 年以来，学校先后派出非洲国家、东南亚国家的校际访问团组，接待了非洲、欧洲等多个高级别访问团，深化了与俄罗斯等国家高等院校的校际关系。二是加大“一带一路”沿线国家留学生招生规模，开设土木工程专业中文授课班，招收来自非洲、南亚等“一带一路”沿线国家和地区的留学生 60 多名。三是举办重点面向“一带一路”的高端学术会议，2016 年学校承办亚洲可持续绿色校园论坛。2017 年下半年，将承办第十届国际供暖通风空调学术会议。四是拓展中外合作办学领域，目前开办了美、英、澳、新等国家的本科专业合作办学项目，以及多所国外大学的学分互认项目。五是深化与“一带一路”沿线国家的科研合作，目前学校与德国、英国、日本、韩国等国家在绿色建筑、建筑工业化等领域开展了深入合作。六是主动对接“一带一路”的人才需求，承办阿尔及利亚等国家的建设高级管理项目培训班，并为中建、中铁、中交等大型国企出国工程人员定制培训班。

我校努力做大做强建筑特色。建筑业是国民经济社会发展的支柱产业，在推进产业转型升级、深化供给侧改革中，发挥着举足轻重的作用。因此，建筑类相关学科专业在今后很长一段时间的经济社会发展乃至社会世界经济大局中，将有很大的发展空间和平台。但也不可否认，从整个世界发展的趋势来看，经济发展程度越高，基础设施建筑方面的产业所占比重会越来越低，而建筑的创意设计、运营维护技术升级、综合应用等所占比重相应越来越高。近几年来，我们学校在巩固传统土建学科优势的基础上，重点发展绿色建筑。全国唯一的绿色建筑技术及其理论博士人才培养项目，就设在我们学校。

我校加速建筑遗产改造、智慧城市等研究方向，成立了山东省绿色低碳建筑研究院，主持国家住建部、能源署的建筑项目——被动式、装配式超低能耗实验楼。目前这个实验楼已经投入运行。最近我们又受住建部委托，举办相关方面的全国培训班。我们的目标是建设立足华东、辐射全国的绿色建筑技术研发成果转换和工程示范中心。

另外，在建筑加固与改造领域，学校也取得了国家技术发明奖、教育部创新团队、山东省泰山学者等重要成果。我们获得国家技术发明奖主要是建筑物移位，目前在建筑物移位这方面，我们逐步地把它延伸做大做强。在这里我们诚邀“一带一路”沿线国家建筑大学共同开展相关领域的科研合作。

关于服务区域发展。我们建筑类高校在“双一流”建设背景下，理应在发展战略上保持清醒，既要主动参与国家“一带一路”开放办学走出去，也要扎根中国大地，扎根区域经济，在服务地方经济、社会发展中，求知识谋发展。近年来学校深入实施“山东建筑大学服务山东建筑”方案，积极搭建强有力的技术研发平台。2016 年，学校绿色建筑与建筑工业化创新实践中心获批为国家级产教融合项目，成为山东省首个获批该项目的高校。学校联合山东师范大学、济南大学等多所高校，共同规划建设齐鲁大学科技园，在青岛“蓝色硅谷”成立绿色建筑创新研究院，在烟台市成立机器人研究院，先后与 40 余家政府和企事业单位签订战略协议，完成技术转化和成果推广 300 余项，在新型城市化建设、建筑工业化、绿色城市、古建筑保护等领域发挥积极作用。

学校土建学科人才集聚，有 8 人担任山东省土木建筑学会城市管理协会、绿色建筑协会、园林协会，房地产协会等会长或理事长，先后牵头起草《山东省产业建筑指导纲要》等政策文件，组织或参与制定了 40 余部国家、省相关行业标准，为建筑事业发展发挥了重要的智囊作用。

各位同仁，“一带一路”倡议给大学的创新发展带来了新的机遇，建筑类大学国际联盟的成立，为学校开展国际交流合作搭建了广阔平台，山东建筑大学将以“一带一路”倡议为纽带，期待着与国内外建筑大学一道，加强互联互通，深化友好交流，开展广泛合作，为“一带一路”沿线国家和地区经济发展作出贡献，欢迎到山东建筑大学指导工作。谢谢大家！

A Report on the Belt and Road Architectural University International Consortium and Presidents' Forum

Fan Cunli, Vice President of Shandong Jianzhu University

Honorable President Zhang Ailin, colleagues, distinguished guests, ladies and gentlemen, good afternoon!

It is a great honor to be entrusted by Professor Jin Fengxiang, the president of Shandong Jianzhu University, to attend the founding meeting of the Belt and Road Architectural University International Consortium initiated and hosted by Beijing University of Civil Engineering and Architecture. After listening to the wonderful speeches made by the presidents of many well-known colleges and universities in architecture at home and abroad, I am very enlightened. Next, I would like to report on some of the basic information about our school.

Shandong Jianzhu University is located in Jinan City, Shandong Province. Shandong Province is known as Qilu Land, the hometown of Confucius and Mencius, which mainly refers to Confucius, the saint in Qufu, Shandong Province, Mencius, the sub-saint in Zoucheng, Shandong Province. Shandong also has Mount Tai, the head of five mountains. And the Yellow River also crosses Shandong Province. Jinan City is known as Spring City, with 72 world famous springs there. Jinan also has a famous Song Dynasty poet Li Qingzhao. Shandong Jianzhu University, founded in 1956, after more than 60 years of development, it has formed a multi-disciplinary university characterized by civil architecture, with engineering as the main subject and interdisciplinary infiltration of engineering, science, management, culture, law, agriculture and arts. We now have 57 undergraduate majors and 60 master degree programs in two disciplines. At present, there are 26,000 students in school, and the annual employment rate of graduates has been in the top three in Shandong Province.

Our school has always attached great importance to the internationalization of running schools. It should be said that internationalization is an important feature of the development of universities in the world today, which exists in different types of universities at different levels, different sizes, different regions and different purposes. Belt and Road Initiative provided opportunities and possibilities for the high-level development of universities in the new century. The establishment of the Belt and Road Architectural University International Consortium, starting from the aspects of school-running system, personnel training, curriculum projects, credit certification, scientific research and so on, and carrying out extensive and comprehensive cooperation with open universities in countries along the Belt and Road Initiative, will certainly promote the internationalization of architectural universities in China. At present, our school has focused on six aspects in promoting internationalized running of schools around the belt and road initiative.

The first is to promote the exchange of high-level visits among countries along Belt and Road Initiative's route. Since last year, our school has sent inter-school visiting groups to African countries and Southeast Asian countries, and received high-level delegations from Africa, Europe, and so on. It has deepened the inter-school relationship with colleges and universities in Russia and other countries.

The second is to increase the enrollment of foreign students from countries along Belt and Road Initiative's route, set up Chinese classes for civil engineering majors and enroll more than 60 foreign students from countries and regions along Belt and Road Initiative, such as Africa and South Asia.

The thired is to hold high-end academic conferences focusing on the belt and road initiative. Our school hosted the Asia Sustainable Green Campus Forum in

2016. The 10th International Conference on Heating, Ventilation and Air Conditioning will be held in the second half of this year.

The fourth is to expand the field of Sino-foreign cooperation in running schools. Currently, there are undergraduate professional cooperation programs in the United States, Britain, Australia, Singapore and other countries, as well as credit mutual recognition programs in many foreign universities.

The fifth is to deepen scientific research cooperation with countries along Belt and Road Initiative. At present, our school has carried out in-depth cooperation with Germany, Britain, Japan, South Korea and other countries in the fields of green building, construction industrialization and so on.

The sixth is to take the initiative to meet the demand for talents in the belt and road initiative, to undertake training courses on construction senior management projects in Algeria and other countries, and to customize training courses for overseas engineers from China Construction, China Railway, China Communications and large state-owned enterprises.

Our school is making an effort to expand and strengthen architectural characteristics. The construction industry is the pillar industry of national economic and social development, which plays an important role in promoting industrial transformation and upgrading and deepening supply-side reform. Therefore, the related major of architecture, in the future for a long time of economic and social development and even the overall situation of the social world economy, will still be a lot of space and platform for development. However, there is no denying the fact that the higher the level of economic development, the lower the proportion of infrastructure construction industry, and the higher the proportion of creative design, operation and maintenance technology upgrading and comprehensive application of buildings. In recent years, based on consolidating the advantages of traditional civil engineering, our school has focused on the development of green buildings. The only national green building technology and its theoretical PhD training program is located in our school.

To speed up the transformation of architectural heritage, smart city and other researches, the Shandong Province Green and Low Carbon Building Research Institute has been established, presiding over the Ministry of National Housing and Construction, the Energy Department Construction Project, passive, assembled ultra-low energy consumption experimental building. At present, this experimental building has been put into operation and recently commissioned by the Ministry of Housing and Construction. Training courses in this area have just been held throughout the country. Our goal is to build a green building technology research and development achievement conversion and engineering demonstration center based in east China and radiating across the country.

Also, in the field of building reinforcement and reconstruction, our school has also won the National Technological Invention Award, the Ministry of Education's Innovation Team, Shandong Taishan Scholars and other important achievements. Our achievement is mainly on building displacement. At present, in the aspect of building displacement, we are going to gradually extend it to become bigger and stronger.

Here we sincerely invite the national architectural universities along the Belt and Road Initiative route to jointly carry out scientific research cooperation in related fields.

Let's talk, about the development of service areas. Under the background of double first-class construction, our architectural colleges and universities should be strategically clear. We should not only take the initiative to participate in the opening up of Belt and Road Initiative to run schools, but also take root in the land of China, and take root in the regional economy. In serving the local economic and social development, we should seek knowledge for development. In recent years, the school has carried out in-depth. Shandong Jianzhu University has served Shandong's construction plan and actively built a strong technology research and development platform. In 2016, the school's Green Building and Building Industrialization Innovation Practice Center was approved as a national-level integration project of industry and education, becoming the first university in Shandong Province to be approved for this project. Our school, in conjunction with Shandong Normal University, Jinan University and other universities, has jointly planned the construction of the Qilu University Science and Technology Park, established the Green Building Innovation Research Institute in Qingdao Blue Silicon Valley and the Robot Research Institute in Yantai City. It has successively signed strategic agreements with more than 40 government enterprises and institutions, completed more than 300 technological transformation and achievements promotion, and played an active role in the new urbanization construction, construction industrialization, green city, ancient architecture protection and other fields.

The school is full of civil engineering talents. Eight of them are the presidents or directors of Shandong Civil Architecture Association's Urban Management Association, Green Architecture Association, Garden Association, Real Estate Association, etc. They have led the drafting of policy documents such as Shandong Industrial Architecture Guidelines, organized or participated in the formulation of more than 40 national and provincial related industry standards, and played an important think tank role for the development of the construction industry.

Colleagues, Belt and Road Initiative has brought new opportunities to the innovation and development of the university. The establishment of the Belt and Road Architectural University International Consortium has set up a broad platform for the school to carry out international exchanges and cooperation. Shandong Jianzhu University will take Belt and Road Initiative as the link, looking forward to working with domestic and foreign architecture universities to strengthen connectivity, deepen friendly exchanges, and carry out extensive cooperation to contribute to the economic development of countries and regions along the line. Welcome to Shandong Jianzhu University to guide the work. Thank you!

在“一带一路”建筑类大学国际联盟校长论坛上的报告

法国英科工程大学副校长　文森特·希克斯

尊敬的校长先生，女士们、先生们，大家下午好！我今天非常荣幸参加本次会议，首先我要特别感谢主办方对我发出的盛情邀请，感谢你们对我们的热情接待。

其实我对于“一带一路”倡议一直保持着浓厚的兴趣，我也非常期待能够在该框架之下，加强各校之间的合作。

首先想和大家做一个介绍，我叫文森特·希克斯，中文的意思是“6”，希望大家记住我。我的职业生涯一开始是结构工程师，主要关注的领域是钢筋混凝土，所以我对本次建筑类大学国际联盟抱有极大的兴趣和极高的希望。下面和大家介绍一下我们的学校。我们学校全称是法国英科工程大学，我主要负责土木工程和建筑学，我曾经做过系主任。在 2015 年的时候，我成为了 HEI 的总经理，除此之外，我也是我们学校负责培训和教学方面的副校长。我们学校可以为大家提供什么呢？我们学校成立于 2016 年，由三个单位组成：法国里尔高级工程师学院（HEI）、法国里尔高等农业学院（ISA）、法国高等电子与数字工程师学院（ISEN），其中 HEI 是 1885 年成立的，历史超过 130 年。现在请大家关注一下这张 PPT，在英科工程大学总学术人数超过 5500 人，毕业生总数超过 27000 人，可以说我们的校友网络是非常强大的。除此之外，在校的教职员工和科研人员超过了 500 人，还有 2500 多个产业界的合作伙伴，并且还和 300 多个国际学术机构开展了合作，有 25 个经过认证的研究实验室。

法国因其卓越的工程质量而闻名，我们的工程学位也是在学生高中毕业之后，要经过5年的专业训练才能授予。想进入法国一流的学术研究机构深造，要经过非常严格的筛选，也正是因为这个原因，我们也算是法国享有盛誉的200多个工程学校之一。除此之外，我们学校每年都有35000多名学生被授予法国工程学位，其中5000多名都是国际学生，法国全国被授予工程学位的学生总数约250万。我们学校的主要使命就是为企业提供支持，为他们提供更好的工程人员和管理人员。每一年我们学校都会有1000多个工程师毕业，其中超过97%的学生都是在毕业之后四个月之内就找到了工作，所有毕业生中，有15%～20%的学生会在海外就业。

所以HEI的一个基本使命就是为工程人员提供高质量训练，提高他们的专业技能，提高他们的责任感，让他们能够胜任重要的技术岗位，并且发挥出团队领导作用，同时具有跨文化、跨学科视角和国际视野。

HEI是在1885年成立的，如今HEI在校生已经达到2400人，HEI会给毕业生授予一个证书，也就是通用工程学位。我们有三个校园，两个在里昂，一个在法国中部。每年我们毕业的工程师人数达到450名，我们也有非常重要的校友网络，大家在PPT上面可以看到毕业生总人数已经达到了16600名。我过去经常跟大家说，我们学校的一大优势就是我们的校友资源、校友构成的网络，而学校开展的三项主要活动，就包括了教学研究和为产业提供的支持。

大家在PPT上面可以看到，我们提供14个主要领域的教学研究，土木工程、建筑设计、力学等，我们也与两所建筑院校开展合作，一个是建筑学校，一个是位于比利时的知名院校，还有智慧城市的相关项目，采用全英文授课。我认为对国际留学生来说，这也是很好的机会，晚些时候会向大家进一步介绍。我们学校教学生如何应对实际中的问题，他们在现场看工程项目，回来做一个相应的模拟，并在这个模拟环境当中进行设计。我们非常注重跨学科训练，课程是全英文授课，能进一步增加学生们的沟通技能，培育他们的国际视野。在教学方面，我们采用合作式、创新式的工作模式，主要就是为了模拟学生们在工作当中建设智慧城市面对的问题。总体上来说，这些培训项目就是为了增加学生的创造和合作精神，这就是我们合作的例子。智慧城市

发展的项目，是我们学校和北京建筑大学合办的项目，欢迎中国的学生来这里学习，我们主要提供两年制的工程学术课程。接下来我放一个视频，当然它是法文语音，但字幕是中文，请大家观赏，大家在视频上可以看到我们HEI的校园。

非常感谢大家！

A Report on the President Forum of the Belt and Road Architectural University International Consortium

Vincent Six, Vice President of Yncréa, France

Dear Mr President, Ladies and Gentlemen,

Good afternoon! I am so honored to attend this meeting today. First of all, I would like to thank the organizers for their kind invitation and thanks for your warm reception.

In fact, I have always maintained a strong interest in the initiative of the Belt and Road. I am also very much looking forward to strengthening cooperation between the schools under this framework.

First of all, I want to introduce to you. My name is Vincent Hicks, which means "six" in Chinese. I hope everyone can remember. My career started as a structural engineer. The main area of my research is reinforced concrete, so I have great interest and high hope for this international consortium. Then I would like to tell you about our school. The full name of the school is called French Yncréa. I am mainly responsible for civil engineering and architecture. I used to be the head of the department and became the general manager of HEI in 2015. In addition, I am also the Vice President of Training and Teaching Law at our school. So what can our school offer to everyone? Then I will give you a brief introduction to our school. Our school was established in 2016 and consisted of three units, namely HEI, ISA, and ISEN. Among them, HEI was established in 1885 and had a history of more

than 130 years. Now, please pay attention to this PowerPoint. The total number of academics in Yncréa is more than 5,500, and the total number of graduates exceeds 27,000. It can be said that our alumni network is very powerful. In addition, there are more than 500 faculty and researchers in the school, more than 2,500 industry partners, and more than 300 international academic institutions, and 25 certified research laboratory.

France is known for its outstanding engineering quality. Our engineering degree is also awarded after five years of professional training after graduating from high school. If you want to enter our French academic research institutions for further study, you must go through a very strict screening. It is for this reason that we are considered to be one of the more than 200 engineering schools in France. There are more than 200 engineering schools in a country. This is not too much. In addition, many students in our school are awarded French engineering degrees every year, more than 35,000 per year, of which more than 5,000 are from international students, while the total number of students awarded in France is 2.5 million. Now, the main mission of our school is to provide support for the company and provide them with better engineers and managers. We graduate more than 1,000 engineers every year, and more than 97% of them are found within four months of graduation. Among all graduates, 15% to 20% of students will be employed overseas.

Therefore, a basic mission of HEI is to provide high-quality training for engineers, improve them professional skills and sense of responsibility, enable them to be competent in important technical positions, and play a leading role in the team, while having an international, cross-cultural, and interdisciplinary perspective. Please see the next page for HEI.

I have just told you that HEI was first established in 1885. Today, HEI has 2,400 students, and HEI will award a certificate, which is a general engineering degree. We have three campuses, two in Lyon and one in central France. We have 450 engineers a year, and we have a very important network of alumni. Everyone can also see the total number of graduates in this area, which has reached 16,600. I used to tell you that the big advantage of our school is the network of our alumni and alumni resources. The three main activities carried out by the school include teaching research and support for the industry.

As you can see from here, we will provide teaching activities in 14 major areas. One is civil engineering, which includes architectural design, mechanics and so on. We have worked with two architecture colleges. One is the bling architectural school and the other is Leroux, a well-known institution in Belgium. In addition, there are related projects in the smart city, which are all taught in English. I think this is a good opportunity for international students to study abroad. Later, we will introduce you further. In addition, we will tell students how to deal with problems in real life. After they see the project on the spot, they will come back to do a corresponding simulation and design in this environment. We also pay great attention to interdisciplinary training, and our courses are taught in English. To further increase students' communication skills and cultivate their international perspective, we adopt a cooperative and innovative working mode in teaching. The main purpose is to simulate the problems faced by students in the construction of smart cities. These training programs, in general, are designed to increase students' creativity and cooperation. This is an example of our cooperation. We can introduce a cooperative degree program to you, which is mainly the support of cooperation between BUCEA and HEI. The project of smart city development is a joint project between our school and the Beijing University of Civil Engineering and Architecture. Chinese students are welcome to come here for training. We mainly provide two years of engineering academic courses. So here is a video to everyone, of course, it is played in French, but the subtitles are Chinese, please watch it.

Everyone can see our HEI campus on the video.

Thank you all!

“一带一路”建设中建筑类高校人才培养策略

河北建筑工程学院校长　师涌江

尊敬的各位领导，各位同仁：

大家好！

2017 年 5 月在北京举行的“一带一路”国际合作高峰论坛，让中国成为世界瞩目的焦点，这是一种新型国际合作模式，以和平合作、开放包容、互学互鉴、互利共赢为指引，以开放为特点，以共商、共享、共建为原则。随着“一带一路”建设的实施，越来越多的中国企业渴望“走出去”，然而高层次国际化复合型人才的匮乏成为最大瓶颈。推进“一带一路”建设，是一项宏大的系统工程，对中国高等教育面向国际市场、服务战略大局、探索新发展职能提出了更高要求，也对我国高等教育国际化发展创造了难得机遇，提供了广阔舞台。高等院校在人才培养、合作办学、人文交流和合作科研等领域大有可为。

高校如何担当起“一带一路”人才培养的战略使命，这成为当前摆在高校面前的一项重要任务。

在新形势下，“服务需求，提高质量”是高等教育深化改革的核心任务，如何才能服务好国家战略需求，是高校迫切需要搞清楚的问题。我们认为，“一带一路”建设，关键取决于人才。推进“一带一路”建设，迫切需要坚实的智力支撑和人才保障。高校要明确自身优势和定位，勇于探索人才培养需求。

“一带一路”建设不仅需要具有广阔视野及开拓精神的领军人才，包括国际化人才和国际人才，还需要基建、制造、生产、运输、安全、金融、经济、外交、法律、语言、历史等各方面的从业人员和管理人员，任何一所大学都难以单独胜任。高校要形成协同合力，尝试不同类型大学的联盟和不同学科的联合，发挥各自优势，建立跨专业、跨学校、跨区域的“一带一路”人才教育平台。

人才培养是高等教育的根本任务，也是推动“一带一路”建设的不竭动力。几年来，国内高校积极探索“一带一路”建设中的人才培养之策，为使拥有千年历史的古老丝绸之路焕发新的生机而持续不断地发力。其实，“一带一路”沿线国家有一个共同特点，就是比较一致地看中了中国的应用型高等教育，特别是工程技术类高等教育，包括理工科建筑类教育，希望中国高校能为当地培育大批应用型技术人才，以更好地对接中国投资，加强对华交流合作，促进当地经济社会快速发展、综合国力显著增强。广大发展中国家普遍面临发展经济的共同使命，希望通过借鉴中国经验，引进中国教育模式与资源，培养当地工程技术人才。

这次由北京建筑大学发起的“‘一带一路’建筑类大学国际联盟”，既契合了这样的需求，也必将为中国高等教育更好地服务“一带一路”建设做出努力和贡献。

（一）以成立机构为载体，对接人才需求

近年来，对接“一带一路”沿线国家的发展战略和人才需求，国内不少高校成立新的机构，并以此为载体，主动合作开展留学生项目。

比如，目前国内许多大学已经与“一带一路”沿线国家开展合作，与中亚国家的一些大学签署联合发展中亚学院的协议，专门设立了中亚留学生奖学金，用于鼓励中亚各国学生前来学习，为这些国家培养应用技术型人才。科学研究是现代大学的重要职能之一，在推进“一带一路”建设中发挥重要作用。比如有的大学与沿线国家大学共同建设“科技与经济研究中心”，建设智库、搭建平台、提供服务，致力于和这些国家共同开展国别与区域研究，推动双方在多方面开展合作，并帮助中资企业培养本土化人才，满足经济社会需求。“一带一路”建设是一项长期的战略任务，既要有高瞻远瞩的战略规划，更要有科学合理的实施方案。“一带一路”建设是一项复杂的系统工程，涉及经济、社会各个方面，高校应积极与有关部门、地方和企业合作，发挥智力优势，可以在

战略研究和项目应用开发领域发挥更加积极的作用，探索“以项目带中心”的发展模式，开展科技与经济交流与合作，同时加大人才培养方面的合作。

（二）以大学联盟为基础，开展全面合作

在服务“一带一路”建设的过程中，众多高校普遍采取组建大学联盟的方式，加强与国内外高校的全面合作。这是一场规模空前的国际教育合作行动。

这次由北京建筑大学首倡发起成立“一带一路”建筑类大学国际联盟，其目的是共同探索跨文化培养与跨境流动的新型人才培养模式，促进联盟高校间师生互动，通过合作办学、专业共建、联合培养、合作创新等方式推动“一带一路”沿线国家建筑类大学的全面交流合作，推动教育综合改革和教育国际化进程，搭建国际化人才培养、科研协同创新及人文交流平台，提高联盟大学的办学活力、人才培养质量、科技创新能力、服务社会能力以及国际交流合作水平；主动适应全球化发展对优秀国际化人才的迫切需求，共同推进“一带一路”各国建设；搭建共商共建共享、促民心相通、文明交流互鉴的智库平台，提升联盟成员在建筑领域的发展及创造力；致力于为人类文明的共融发展和高等教育的开放合作贡献力量，以加强联盟大学的全面交流与合作，推动教育综合改革和教育国际化进程，搭建国际化人才培养、科研协同创新及人文交流平台。

“一带一路”建筑类大学国际联盟将致力于高素质、国际化建筑类工程技术人才培养、培育模式的创新实践，为企业培养高水平、实践能力强的建筑类工程技术人才。

（三）以学科优势为依托，培养专业人才

作为工科学科为主、建筑特色鲜明、优势专业突出的建筑类大学，应积极发挥优势，主动布局，以学科优势为依托，加强专业人才的培养，为“一带一路”沿线国家培养中国化的建筑类工程人才。

一是有条件的大学应主动采取措施，尽早谋划，立即布局、调整留学生工作，重点搭建覆盖“一带一路”沿线国家的高层次招生平台。二是重点加

强建筑工程类全英文授课专业建设，培养更多工程领域的高水平教师，并长期参与留学生的授课和培养。三是以建筑企业拓展海外业务而开展的海外合作项目为依托，联合企业，开展建筑类专业方面的国际化订单式人才培养。四是着力构建“一带一路”国家教育共同体合作平台，主动与沿线国家的高校、企业开展合作，并与国内企业签署培养来华留学生的合作协议，为企业培养当地人才，解决人才短缺尤其是技术管理人才短缺的问题。

（四）拓展人文交流，发挥骨干作用

人文交流在“一带一路”建设中承担着不可替代的重要职能。高校拥有文化和知识优势，是当代社会重要的知识生产和传播基地，是开展人文交流的主体力量。高校面向“一带一路”沿线地区，因地制宜，广泛开展各种形式的校际合作、师生交流、合作研究，文化、艺术、体育交流，既可以推动知识传播，促进地区发展，也可以推动文明交流，促进人心相知，营造良好国际环境。同时，高校面向“一带一路”沿线各地广泛开展各种形式的教育交流合作，深化人文交流，也是学校国际化发展的重要价值取向之一和具体体现。目前，我国高校与“一带一路”沿线国家的教育交流合作，总体上还不够深入，人文交流还不够广泛。有影响力的地区教育合作机制和人文交流机制不多，交流渠道不够畅通。尤其是建筑类高校关于“一带一路”沿线的知识储备和师生认知水平总体上还比较低，专门研究“一带一路”沿线国家的人员屈指可数，图书资料也不多。这就意味着建筑类高校在深化与“一带一路”沿线地区人文交流方面大有可为。

（五）抢抓发展机遇，服务区域建设

张家口是多民族之间交流的古道，也是欧亚大陆中重要的国际贸易口岸城市。当前，张家口正在全力借助冬奥会和“一带一路”“草原丝绸之路”的世界影响力，致力于搭建张家口走向世界和让世界了解张家口的桥梁。

近年来，张家口面临抢抓京津冀协同发展、京张携手举办冬奥会、建设国家可再生能源示范区三大机遇。京津冀协同发展和“一带一路”建设作为重大国家战略，为河北省高等教育和河北建筑工程学院的发展带来了重大历

史机遇，创造了极为有利的发展环境。

一是学校将主动融入京津冀，在推进学校综合改革、促进教育资源共建共享、推动协同创新等方面发挥示范引领作用，为京津冀协同发展提供智力、技术、人才等方面的有效支撑，在建筑领域发挥不可替代的作用。二是抢抓京张联合举办 2022 年冬奥会的有利时机，发挥学校是张家口地区唯一一所工科本科院校的办学优势和专业特点，利用一切可能的机会，支持奥运、服务奥运，在规划设计、场馆建设、市政工程、环境保护、装备保障、志愿服务、人才培养等方面发挥积极作用。三是学校作为区域科技资源的主要聚集地，主动契合张家口市可再生能源示范区建设的需求，进一步强化社会服务意识，为可再生能源示范区建设提供战略决策咨询支持。四是学校将置身“一带一路”建设，加快建设以建筑产业类学科为特色的多科性大学，积极推进建筑节能和绿色建筑发展，积极发挥自身学科优势对提升行业发展水平的支撑和引领作用，培养一大批能站在多领域技术发展前沿、具有综合能力的现代工程技术和管理人才，在服务国家战略和区域经济社会发展中凸显学校地位。

总之，“一带一路”的建设将是一个长期而艰巨的过程，迫切需要坚实的智力支撑和人才保障，为国内高等院校创造了新的发展机遇，提出了新的更高要求，要求高校加快教师队伍建设和教育教学改革，加快提升国际化发展水平，努力发挥更加积极的作用。我们相信，乘着“一带一路”国际合作高峰论坛的东风，“一带一路”建筑类大学国际联盟必将为培养建筑类国际人才贡献更大的力量。

谢谢！

Training Strategies of Talents in Architectural Universities While Building the Belt and Road

Shi Yongjiang, President of Hebei University of Architecture

Honorable leaders and colleagues:

How are you!

The Belt and Road Forum of International Cooperation which was held in May, 2017 made China become the focus of world's attention. This is a new international cooperation model which is guided by peaceful cooperation, openness and inclusiveness, mutual learning and reference and mutual benefits and win-win result. It is characterized as openness and also adheres to the principles of discussion together, sharing and building together. More and more Chinese enterprises are eager to go abroad to conduct business after the implementation of the Belt and Road strategy. But the lack of high level international inter-disciplinary talents has become the largest bottleneck. Promoting the build of the Belt and Road is a grand systematic project and it leads to higher requirements for Chinese higher education to face international market, serve the overall strategic situation and explore new development functions. It also provides us with valuable opportunities and a broader stage for internationalization and development of higher education. Higher education institutions make contributions to training of talents, cooperation in running schools, cultural exchange and cooperation in scientific research and so on.

How do colleges and universities take the responsibility for the training of the

talents for the Belt and Road? It has become a current important mission for higher education institutions.

Under the new circumstance, "Serve for the needs, improve the quality" is the core mission for the deepened reformation of higher education. And how to serve well for the needs of national strategies is the problem needs to be solved for colleges and universities. We believe that talents play important roles in the establishment of the "Belt and Road". Promoting the establishment of the "Belt and Road" requires urgently the strong support of intelligence and guarantee of talents. Colleges and universities should ensure their advantages and location while exploring the requirements for training of talents.

Not only does the establishment of the "Belt and Road" need leaders who own broad horizon and pioneering spirit including international talents, but also need the employees and managers who are specialized in infrastructure, manufacture, production, transportation, safety, finance, economics, diplomacy, law, language, history and so on. And any university is unable to be qualified alone. Colleges and universities should cooperate and try the alliance of different types of universities and the combination of different disciplines and make use of respective advantages to establish the educational platform for training of inter-disciplinary, inter-university and inter-region talents for the "Belt and Road".

Training talents is the essential mission for higher education and the lasting impetus of the promotion for the establishment of the "Belt and Road". In the last few years, domestic colleges and universities have made an effort to explore the strategies of training talents for the establishment of the "Belt and Road" in order to keep on promoting the ancient Silk Road with a history of thousands of years to display new vigour. In fact, the countries along the "Belt and Road" all have a common characteristic. It is that all of them think highly of the practical higher education in China consistently, especially the higher education for engineering technologies including the education for architecture in science and engineering field. And they hope that Chinese colleges and universities can train lots of practical and technical talents for local in order to realize better butt joint with Chinese investment and strengthen the communication and cooperation with China and promote the fast development of local economic society and comprehensive national strength. A lot of developing countries usually have the common mission to develop economics. By referring to Chinese experience and introducing Chinese educational models and sources, they can train local engineering technical talents.

The "Belt and Road Architectural University International Consortium" which is initiated by Beijing University of Civil Engineering and Architecture meets the requirements and it also serves to the Chinese higher education better while making contribution to the strategies of "Belt and Road" .

1. Establishment of institutions is supporter to the needs of docking talents

In the last few years, to dock the developing strategies and needs of talents for the countries which are along the "Belt and Road", many domestic colleges and universities establish new institutions and use these institutions as supporters to initiate forwardly projects related to overseas students.

For example, at present, many domestic universities have signed agreements about jointly developing central Asian institutions with countries along the "Belt and Road", such as some central Asian countries. Some scholarships for central Asian overseas students are established specially to attract the students from central Asian countries to study here and train practical and technical talents for these countries. Scientific research is one of the important functions of modern universities and it plays an important role in promoting the establishment of the "Belt and Road". For example, some universities are cooperating with countries along the "Belt and Road" to establish "Science and economic research center" . They will establish think tanks, set up platforms and offer services to devote themselves to research countries and regions with these countries, promote mutual cooperation in many aspects and help China-invested enterprises to train local talents to satisfy economic and social needs. The establishment of the "Belt and Road" is a long-term strategic mission which needs not only farsighted strategic plans but also reasonable scientific

implementation schemes. The establishment of "Belt and Road" is a complicated systematic project which involves many aspects of economics and society. Colleges and universities should cooperate with related departments, local partners and enterprises. Using the advantage of intelligence can play a positive role in research of strategy and development of the projects. Developing model in which the projects are core should be explored. Communication and cooperation about science and economics should be undertaken while strengthening the cooperation about training talents.

2. Carry out overall cooperation based on university consortium

Many universities adopt the method of establishing university consortiums to strengthen the overall cooperation with domestic and foreign universities while serving for the "Belt and Road" significant strategy. And this is an international education cooperation activity on an unprecedented scale.

The aim of Belt and Road Architectural University International Consortium, which was initiated by Beijing University of Civil Engineering and Architecture, is to explore the new talents training model of cross-cultural training and cross-border flow and to promote the interaction between teachers and students in universities of consortium. By cooperation in education, co-construction of speciality, joint training, cooperation in innovation and so on to promote the overall cooperation between the Belt and Road Architectural Universities. We should promote the synthetical reformation in education and internationalization of education, establish the platform for training of international talents, collaborative innovation in science and cultural communication. We should improve the educational vitality of universities in consortium, quality of talents training, capability of scientific innovation, capability of serving the society and level of international communication. We should forwardly adapt to the imperious needs of excellent international talents for globalized development and establish think tank platforms for joint business, joint establishment, sharing, promotion of communication of popular feelings, communication and reference of culture. We should improve the development and creativity in the field of architecture. We should devote ourselves to making contributions to the development and integration of human culture and openness and cooperation of higher education. We should enhance the overall communication and cooperation of universities in consortium and promote the synthesized reformation of education and internalization of education. We should establish the platform for training of international talents, collaborative innovation in science and cultural communication.

The "Belt and Road Architectural University International Consortium" will devote itself to training of high-quality and international architectural engineering technological talents, innovation of educational model, training high-level architectural engineering technological talents with strong practical ability.

3. Train professional talents based on the advantages of subjects

As an architectural university with highlighted dominant majors and distinctive architectures which focuses on engineering subjects, it should exert its advantages positively and strengthen the training of professional talents based on the advantages of subjects in order to train "Chinese" architectural engineering technological talents for countries along the "Belt and Road".

Firstly, universities satisfying the conditions should take measures positively and make plans in advance to arrange and adjust the matters about overseas students and focus on establishing high-level enrollment platforms covering the countries along the "Belt and Road". Secondly, we should focus on strengthening the professional establishment of all-English architectural engineering lectures in order to train more high-level teaches in the field of engineering who take part in the training and lectures of overseas students. Thirdly, based on overseas cooperation projects whose aims are to expand the overseas business for architectural enterprises, we should cooperate with enterprises to carry out international and customized training of the talents in the field of architecture. Fourthly, we should make effort to construct cooperation platforms for the education of countries along the "Belt and Road" and cooperate with countries and enterprises along the "Belt and Road". And we should sign cooperation agreements with domestic enterprises about training overseas students who come to China for studying to train local talents for these enterprises

and solve the problem of lacking talents, especially the issue about the shortage of technological management talents.

4. Play a key role and expand communication about culture

Communication about culture plays an irreplaceable and important role in the construction of "Belt and Road". Colleges and universities, which are important bases of production and diffusion of knowledge and are main force for people-to-people communication, have advantages about culture and knowledge. Considering the regions along the "Belt and Road", colleges and universities adjust measure to local conditions to carry out different types of cross-universities cooperation, communication between teachers and students, study and cooperation. Communication about culture, art and sports can promote the distribution of knowledge and development of region. It also can promote the communication about culture and meeting of people while building good international environment. Meanwhile, considering regions along the "Belt and Road", colleges and universities carry out many types of communications and cooperation about education. And deepening communication of culture is also one of the important value orientations and concrete indications of universities'international development. At present, the communication and cooperation between domestic universities and countries along the "Belt and Road" is not thorough enough and communication about culture is not broad enough. There are not enough influential regional education cooperation and cultural exchange mechanism. Most of all, architectural universities do not have both high-level knowledge storage about "Belt and Road" and high cognitive level of teachers and students. People who are specialized in the research about countries along "Belt and Road" are few and the related books and materials are also few. It means that we can make great contribution to deepening communication between architectural universities and regions along the "Belt and Road".

5. Seize the development opportunities and serve the construction of regions

Zhangjiakou is an ancient channel where people in multi nations communicate with each other while being important international trade port city in Eurasia. At present, relying on the world influence of Olympic Winter Games, "Belt and Road" and "Prairie Silk Roads", Zhangjiakou is trying hard to build the path for leading Zhangjiakou to the world and make the world know more about Zhangjiakou.

In the last few years, Zhangjiakou City has met three opportunities including coordinated development in Beijing-Tianjin-Hebei Region, cooperation between Zhang and Beijing about holding Olympic Winter Games and constructing national demonstration area of renewable energy sources. As important national strategies, coordinated development in Beijing-Tianjin-Hebei Region and construction of "Belt and Road" provide higher education in Hebei and Hebei University of Architecture with historic opportunities of developing while creating favourable developing environment.

Firstly, universities will positively be a part of Beijing-Tianjin-Hebei Region and play a leading role in promoting synthesized reformation of universities, co-construction and sharing educational sources, cooperation, innovation and so on. They will offer effective supports about intelligence, technologies and talents, who play irreplaceable roles in the field of architecture, for development of Beijing-Tianjin-Hebei Region. Secondly, we should seize the opportunity of holding the Olympic Winter Games in 2022 by Beijing and Zhangjiakou. We should also make use of the advantages and major characteristics of the only engineering college located in Zhangjiakou region to support the Olympic Games and serve the Olympic Games by using all possible chances. We should play positive roles in planning and designing, construction of stadiums, municipal projects, environmental protection, equipment support, voluntary services, training the talents and so on. Thirdly, as the main gathering place of regional scientific resources, the university will positively meet the needs of the construction of demonstration area for renewable energy source in Zhangjiakou city. Ant will further strengthen the spirit of serving society to offer strategic decision-making consultant support to demonstration area for renewable energy source. Fourthly, the university will place itself in the national strategies such as the

construction of "Belt and Road" in order to accelerate the construction of multi-disciplinary university which is characterized by architectural industrial subjects. The university will promote the energy-saving in the architectures and development of green buildings. It will also make use of its disciplinary advantages which own supporting and leading impact on improving the level of the industry. The university will train a lot of modern engineering and technological management talents who stand on the front line of development of multi-field technologies.

Above all, the "Belt and Road" will be a long-term and onerous process which requires massy intelligent support and guarantee of talents. It has created new opportunity for the development of domestic colleges and universities and puts forward to new higher demands for which colleges and universities should accelerate the construction of teachers'teams and reformation of education, accelerate the improvement of international level and try to play a better role in that. We believe, relying on the Belt and Road Forum of International Cooperation, the "Belt and Road Architectural University International Consortium" will definitely make greater contributions to training international architectural talents.

Thank you very much!

在“一带一路”建筑类大学国际联盟校长论坛上的报告

马来西亚大学联盟主席　拿督斯里·黄子炜
马来西亚理工大学首席执行官　卡西姆

黄子炜：大家下午好！感谢大会给我这个机会，下面我把时间交给马来西亚理工大学首席执行官卡西姆。

卡西姆：感谢黄博士。首先，我先要祝贺北京建筑大学能够成功地召开这次“一带一路”建筑类大学国际联盟论坛；其次，我要感谢这次邀请，让我们加入到联盟之中；此外，还要感谢黄博士的邀请。

我们大学一共有两个校园，其中一个是在吉隆坡城内，我们的主校区则位于马来西亚最南端的城市，是一个非常大的校园，面积约 1148 平方千米。PPT 上是我们校园的景观，前三张图片是我们位于南端的大主校区，第四张图片展现的是吉隆坡的校区。

我们一共有 21000 多名学生，其中 9000 名是研究生，博士生和硕士生比例约为 1∶1。到 2020 年时，我们预计，本科生和研究生的比例会达到 4∶6。我们目前有 2700 多名国际留学生，对于我们联盟来说，也是具有重要意义的，因为我们的学生，不论是境外留学的，还是我们本土就读的，学生数量都非常多，我们也希望能够借助联盟来吸引更多的学生。

在马来西亚理工大学，教职员工各司其职。总体来说我们有 55 个本科项目，82 个硕士学位的课程，还有 120 个能授予博士学位的专业研究项目，所以我们在这边研究生和博士研究生比例相当之高，建筑学也是我们的重点学科。我们大学会在一些新的领域进行创新，比如前沿科学、创新性工程、公共健康、智慧数字社区发展，还有资源可持续发展。刚才我们巴黎的同事谈到了智慧城市发展的重要意义，我们也完全赞同。在这些领域中，我们有不同的研究所，有专门的研究项目（或者称之为卓越中心），都会发表相应的论文和学术出版物。说到国际化进程，我们有 62 个不同国家和地区的学生，主要来自东盟国家，如南亚、东亚、中东、北非等，还有孟加拉国、巴勒斯坦和新加坡的国际留学生，这些都是我们生源的重要组成部分，所以我们有着数量众多的留学生，这也体现出了我们的国际性。我们在全球大学排名当中综合排名 21 位，除此之外，我们电力工程，以及所有的工程性学科、环境和建筑方面，排名都进入了全球高校学科百强。我认为通过我们和联盟的合作，可以在未来进一步提升我们的排名，希望在 2020 年，至少环境和建筑专业能够跻身全球 50 强。我们全球的合作学术研究项目主要有 300 多个，合作对象包括全球知名的麻省理工学院，以及其他的全球知名院校。我们一直以来都非常期待和杰出的机构开展合作；即使有的机构的学术能力并不是特别强，我们也很愿意和他们建立合作关系。换言之，我们愿意和所有人进行合作，和所有人共同成长。

接下来想跟大家介绍的是，我们希望在“一带一路”建筑类大学国际联盟中，找到新的优势领域。我们有了东盟建筑类大学的优势，也有了一些全球合作优势，在“一带一路”建筑类大学国际联盟的基础上，我们可以汇聚之前的优势。刚才跟大家说过了，我们和麻省理工学院开展了合作，这个合作对我们来说非常重要，合作项目主要授予研究生类的学位。我们和麻省理工学院联合进行学生的筛选，学生被选中之后，会在我们大学学习 5 个月，在麻省理工学院学习 4 个月，在博士后项目学习 9 个月，再在两地进行研究。我们这个项目做了 5 年，未来还要继续 5 年。

再接下来介绍几个我比较熟悉的部门。这个部门叫做建筑环境学院，由 1994 年建立的建筑学系、1974 年建立的量化勘测学系、1974 年建立的城市与区域规划学系合并而成。2002 年我们增设了一个景观项目。除此之外还有两个研究中心，其中一个叫门类建筑中心，主要是研究整个马来西亚的建筑文化遗产，还有一个是专注于城市市政总体规划的发展研究中心。这几个部门教职员工有 28 个，77% 的教职员工都拥有博士学位。我们大学总教职员工达到101人。

这页PPT展示了我们的招生情况，刚才说到的四个部系，其中有5个本科项目和7个研究生项目，现学生总数是1310人。我们学校的研究生比例相对比较高，在整个马来西亚大学联盟中，研究生比例只占到32%，68%都是本科生。

下面这页PPT介绍了我们给“一带一路”建筑类大学国际联盟带来的优势。我们学生的流动性比较好，可以和大家开展学生活动。我们是很好的留学生接收地，也是留学生的生源地。我们有一个夏令营项目，学生在夏令营中可以待一到两个学期，现在我们已经进入新学期，如果大家感兴趣的话，可以在这个学期当中，派留学生到我们这边进行进一步的深造。

大家可以看到，PPT上是我们截至去年的国际留学生交流的统计数据；大家也可以看到，无论是出境学习，还是到我们大学学习的国际留学生的数量还是比较多的。除此之外，我们也会组织各种各样的学术研讨会和交流活动，比如2017年我们举办的全球工程教育论坛，而且在2018年和2019年都会有一些非常重要的教育活动，我们是主办方。我说的内容就这么多，现在把麦克风再次交给黄子炜先生。

黄子炜：我们这个大会，明年将会在这个大学主办。谢谢大家！

Report at Belt and Road Architectural University International Consortium Presidents' Forum

Dato'Seri Wong Tze Wei, President of Consortium of Malaysian University; Salleh Bin Kassim, CEO of University of Technology Malaysia

Wong Tze Wei: Good afternoon, everyone!

First of all, I would like to thank the conference for giving me the opportunity to say a few words here. I will leave my time to the local University of Technology Malaysia to introduce this university.

Kassim: Thank you, Dr. Wong. First of all, I would like to congratulate the Beijing University of Civil Engineering and Architecture on the successful convening of this Inaugural Meeting of Belt and Road Architectural University International Consortium. Today, we also want to thank Dr. Wong for inviting us to join this consortium.

This is the location of our university. We have two campuses. One is in Kuala Lumpur. Our main campus is in the southernmost city of Malaysia. It's a very large campus with almost 1,148 square kilometers. This is the view of our campus. The first three pictures are our large main campus located in the south end, and the fourth picture shows the campus site in Kuala Lumpur.

We have a total of more than 21,000 students, of whom 9,000 are postgraduates, almost half of them are with a doctorate and half are with a master degree. By 2020, we expect that the proportion of undergraduates and postgraduates will be almost 4 to 6. We currently have about 2,700 international students. For our consortium, it's also of great significance, because the scale of our students, both those who go abroad to study and those who come here to study, is very large. We also hope to attract more students through consortium.

In our University of Technology Malaysia, there are faculty and staff in various departments. For example, there are staffs in different departments on the campus. In addition, we also have campuses in other branch campuses. Generally speaking, we have 55 undergraduate programs, moreover, there are 82 master's degree courses and 120 professional research projects awarded doctoral degrees, so the proportion of postgraduates and doctoral candidates here is quite high. Architecture is basically our key discipline. In our university, we will find some new fields, such as frontier science, innovative engineering, public health, smart digital community development, and sustainable development of resources. Because our colleague in Paris just talked to us about the significance of the development of smart cities, we also fully agree with them here. Another focus, as I just talked to you, about the sustainability of resources. In these fields, we have different research institutes, and special research projects, or call it a center of excellence, and we will publish corresponding papers and academic publications. When it comes to our internationalization process, we have students from 62 different countries and regions, and there are mainly 15 countries from ASEAN, South Asia, East Asia, Middle East and North Africa and so on. For example, international students from Bangladesh, as well as from Palestine and Singapore, are all important components of us. Therefore, we have a large scale or a large number of international students on campus, which also reflects our internationalism. Here, you can see our various global rankings. We ranked the 21st in the comprehensive rankings of global universities. In addition, we ranked among the top 100 in electric power engineering, as well as all engineering disciplines, environment and architecture. I think it's also a way for us to further strengthen our ranking in the future so

that by 2020, at least in terms of environment and architecture, we can be among the top 50 in the world through our cooperation with the consortium. In this regard, I would like to tell you that there are mainly more than 300 academic research projects for our global cooperation, including MIT, which ranks first in the world, and other well-known global institutions. We have always been looking forward to cooperating with outstanding institutions. However, even if the academic ability of some institutions is not particularly strong, we are also willing to establish cooperative relations with them. In other words, we are willing to cooperate with everyone and grow together with everyone.

After this, I would like to introduce to you that we also hope that we can find our new field of expertise on based on our architecture under the context of The Belt and Road initiative (namely our international consortium). Because now we have the advantages of ASEAN's architectural universities, and we have all kinds of other advantages in the world, then we can gather the previous advantages on the basis of the consortium of architectural universities. As I said to you just now, we cooperated with MIT. This project can be said to be very important for us. It mainly awards postgraduate degrees to countries with South-South cooperation and the Group of 77. We work with MIT to select students. After students are selected, they will stay for five months to study. They will stay for four months at MIT in Boston and 9 months for postdoctoral projects, researching both places. We have been working on this project for five years and will continue for another five years.

In this part, it is also a department I am familiar with. This college is called the School of the Built Environment, which includes the Department of Architecture established in 1994, the Department of Quantitative Survey established in 1974 and the Department of Urban and Regional Planning established in 1974. In 2002, we added a landscape project. In addition, there are two research centers, one of which is called the category architecture center, which mainly stu-dies the architectural cultural heritage of the whole Malaysia. After that, we also have another development research center focusing on the overall urban planning. Let me introduce our faculty and staff. We have 28 academic staff here, including associate professors and professors. Of all the faculty and staff teams, 77% of them have PhD degrees. We have a total of 101 staff.

On this page, the figures we see are about our enrollment, which only refers to the four departments I just mentioned. We currently have a total of 1,310 students, and there are mainly five undergraduate programs and seven graduate programs. And although I mentioned that the proportion of graduate students in our university is relatively high, in our entire college, the proportion of graduate students only accounts for 32%, and 68% are undergraduates.

This page describes what advantages we will bring to the consortium. First of all, our students have good mobility. For example, we can carry out student activities with you. This mainly includes two kinds: one is that we are a very good place to receive overseas students, and we are also the source of overseas students. Specifically, we say that we are the receiving place of overseas students because we have a summer camp program, in which students can stay one or two semesters. Now we have entered a new semester beginning in November. If you are interested in it, you can send overseas students to our university for further study in this semester.

Here, you can see our statistics on international student exchanges as of last year. As you can see, the number of international students who go abroad to study or those who come to our university to study is quite large. In addition, we will also organize various academic seminars and exchange activities. For example, we will soon hold global engineering education forum in 2017, and there will be some very important educational activities in 2018 and 2019, which will be hosted by us. That's all I want to say. Now let's give the microphone to Mr. Wong Tze Wei again.

Wong Tze Wei: Next year, our conference will be held in this university. Thank you!